Rainer Mallée, Werner Fuchs,
Rolf Eligehausen

Design of Fastenings for Use in Concrete – the CEN/TS 1992-4 Provisions

Related Titles

Since it was founded in 1906, the Ernst & Sohn "Beton-Kalender" has been supporting developments in reinforced and prestressed concrete. The aim was to publish a yearbook to reflect progress in "ferro-concrete" structures until – as the book's first editor, Fritz von Emperger (1862-1942), expressed it – the "tempestuous development" in this form of construction came to an end. However, the "Beton-Kalender" quickly became the chosen work of reference for civil and structural engineers, and apart from the years 1945-1950 has been published annually ever since.

approx. 230 pages,
Softcover.
approx. € 49,90

ISBN: 978-3-433-03041-7
Publication date:
approx. June 2013

JÜRGEN GRÜNBERG,
JOACHIM GÖHLMANN

Concrete Structures for Wind Turbines
Beton-Kalender Series
Series editor: K. Bergmeister, F. Fingerloos, J.-D. Wörner (eds.)

The wind energy industry in Germany has an excellent global standing when it comes to the development and construction of wind turbines. Germany currently represents the world's largest market for wind energy.

approx. 210 pages,
Softcover.
approx. € 49,90

ISBN: 978-3-433-03043-1
Publication date:
approx. September 2013

BERND HILLEMEIER,
CLAUS FLOHRER,
JÜRGEN KRELL,
GABRIELE MARQUARDT,
JEANETTE ORLOWSKY,
MICHAEL RAUPACH,
KARSTEN SCHUBERT,
STEPHANIE SCHULER

Reconditioning and Maintenance of Concrete Structures
Beton-Kalender Series
Series editor: K. Bergmeister, F. Fingerloos, J.-D. Wörner (eds.)

The rehabilitation and maintenance of concrete structures takes place in the tense atmosphere of the different interests of client, building authorities, construction companies and material manufacturer.

approx. 144 pages,
70 fig., 14 tab.,
Softcover.
approx. € 49,90

ISBN: 978-3-433-03042-4
Publication date:
approx. April 2013

RÜDIGER MEISWINKEL,
JULIAN MEYER,
JÜRGEN SCHNELL

Design and Construction of Nuclear Power Plants
Beton-Kalender Series
Series editor: K. Bergmeister, F. Fingerloos, J.-D. Wörner (eds.)

Despite all the efforts being put into expanding renewable energy sources, large-scale power stations will be essential as part of a reliable energy supply strategy for a longer period. Given that they are low on CO2 emissions, many countries are moving into or expanding nuclear energy to cover their baseload supply.

approx. 130 pages,
81 fig., 4 tab.,
Softcover.
approx. € 49,90

ISBN: 978-3-433-03044-8
Publication date:
approx. April 2013

RAINER MALLÉE,
WERNER FUCHS,
ROLF ELIGEHAUSEN

Design of Fastenings for Use in Concrete – the CEN/TS 1992-4 Provisions
Beton-Kalender Series
Series editor: K. Bergmeister, F. Fingerloos, J.-D. Wörner (eds.)

The European pre-standard CEN/TS 1992-4 for the design of fastening by means of headed studs, anchor channels as well as post-installed mechanical and chemical anchors is ready for use.

Online-Order: www.ernst-und-sohn.de/books

Ernst & Sohn
A Wiley Company

Ernst & Sohn
Verlag für Architektur und technische Wissenschaften GmbH & Co. KG

Customer Service: Wiley-VCH
Boschstraße 12
D-69469 Weinheim

Tel. +49 (0)6201 606-400
Fax +49 (0)6201 606-184
service@wiley-vch.de

*€ Prices are valid in Germany, exclusively, and subject to alterations. Prices incl. VAT. Books excl. shipping. Journals incl. shipping. 0257300006_dp

Rainer Mallée, Werner Fuchs,
Rolf Eligehausen

Design of Fastenings for Use in Concrete – the CEN/TS 1992-4 Provisions

Ernst & Sohn
A Wiley Brand

The Authors

Dr.-Ing. Rainer Mallée
Stockengartenstr. 12
72178 Waldachtal
Germany

Dr.-Ing. Werner Fuchs
University of Stuttgart
Institute of Construction Materials
Pfaffenwaldring 4
70569 Stuttgart
Germany

Prof. Dr.-Ing. Rolf Eligehausen
University of Stuttgart
Institute of Construction Materials
Pfaffenwaldring 4
70569 Stuttgart
Germany

The Editors of Beton-Kalender

Prof. Dipl.-Ing. DDr. Dr.-Ing. E. h. Konrad Bergmeister
Ingenieurbüro Bergmeister
Peter-Jordan-Str. 113
1180 Wien
Austria

Dr.-Ing. Frank Fingerloos
German Society for Concrete and Construction Technology
Kurfürstenstr. 129
10785 Berlin
Germany

Prof. Dr.-Ing. Dr. h. c. mult. Johann-Dietrich Wörner
German Aerospace Center
Linder Höhe
51145 Köln
Germany

Translation: *Dr.-Ing. Werner Fuchs,*
Stuttgart, Germany

Cover: City-tunnel Malmö, Sweden
(2005-2009)
© fischer Sverige AB
Niklas Svensson

The original German text is published in Beton-Kalender 2012 (ISBN 978-3-433-02989-3) and titled: *Bemessung von Verankerungen in Beton nach CEN/TS 1992-4.*

Library of Congress Card No.: applied for

British Library Cataloguing-in-Publication Data
A catalogue record for this book is available from the British Library.

Bibliographic information published by the Deutsche Nationalbibliothek
The Deutsche Nationalbibliothek lists this publication in the Deutsche Nationalbibliografie; detailed bibliographic data are available on the Internet at <http://dnb.d-nb.de>.

Print ISBN: 978-3-433-03044-8
ePDF ISBN: 978-3-433-60270-6
ePub ISBN: 978-3-433-60271-3
mobi ISBN: 978-3-433-60272-0
oBook ISBN: 978-3-433-60273-7

Typesetting Thomson Digital, Noida, India
Printing and Binding betz-druck, Darmstadt, Germany

Printed on acid-free paper.

Contents

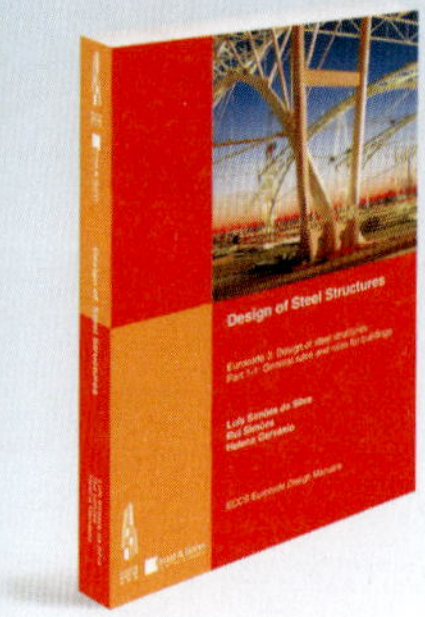

ED.: ECCS

Design of Steel Structures
EC3: Design of Steel Structures. Part 1-1: General Rules and Rules for Buildings.
2010.
446 pages, 295 fig., 105 tab., Softcover.
€ 70,–*

ISBN 978-3-433-02973-2

■ This book introduces the basis design concept of Eurocode 3 for current steel structures, and their practical application. Numerous worked examples will facilitate the acceptance of the code and provide for a smooth transition from earlier national codes to the Eurocode.

ED.: ECCS

Fire Design of Steel Structures
EC1: Actions on Structures. Part 1-2: Actions on Structures exposed to Fire.
EC3: Design of Steel Structures. Part 1-2: Structural Fire Design.
2010.
428 pages, 134 fig., 21 tab., Softcover.
€ 70,–*

ISBN 978-3-433-02974-9

■ This publication sets out the design process in a logical manner giving practical and helpful advice and easy to follow worked examples that will allow designers to exploit the benefits of the new approach given in the Eurocodes to fire design.

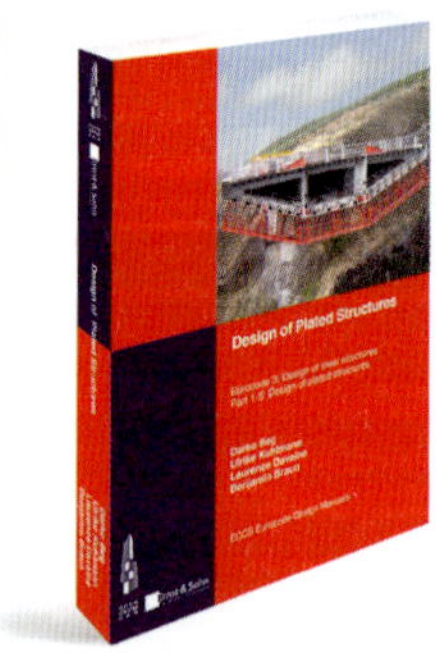

ED.: ECCS

Design of Plated Structures
EC3: Design of Steel Structures. Part 1-5: Design of Plated Structures.
2011.
272 pages,100 fig., Softcover.
€ 55,–*

ISBN 978-3-433-02980-0

■ This design manual provides practical advice to designers of plated structures for correct and efficient application of EN 1993-1-5 design rules and includes numerous design examples.

The ECCS – The European Convention for Constructional Steelwork – unites all of the European steel construction associations, which have global connections to steel construction companies, architects, planning offices, universities and research institutes under one roof. This combined effort also results in publications for the international market by authors who are making crucial contributions to steel construction development. In future, these books will be jointly published with the Ernst & Sohn publishing company.

ED.: ECCS

Fatigue Design of Steel and Composite Structures
EC3: Design of Steel Structures. Part 1-9: Fatigue.
EC4: Design of Composite Steel and Concrete Structures.
2011.
311 pages, 103 fig. Softcover
€ 55,–*

ISBN 978-3-433-02981-7

■ This volume addresses the specific subject of fatigue, a subject not familiar to many engineers, but still relevant for proper and good design of numerous steel structures. It explains all issues related to the subject: Basis of fatigue design, reliability and various verification formats, determination of stresses and stress ranges, fatigue strength, application range and limitations. It contains detailed examples of application of the concepts, computation methods and verifications.

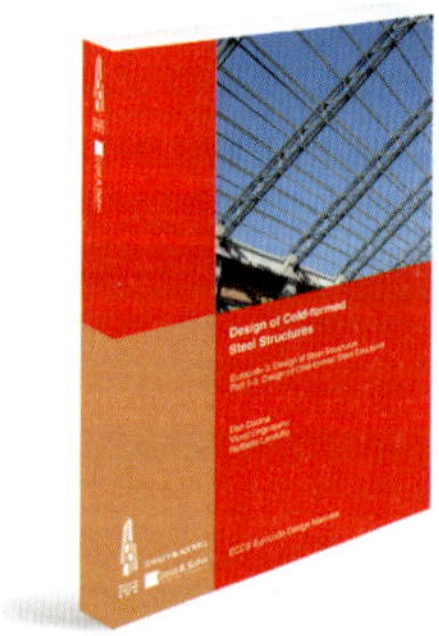

ED.: ECCS

Design of Cold-formed Steel Structures
EC3: Design of Steel Structures. Part 1-3: Design of Cold-formed Steel Structures.
October 2012.
512 pages, 300 fig., Softcover.
€ 70,–*

ISBN 978-3-433-02979-4

■ The book is concerned with design of cold-formed steel structures in building based on the Eurocode 3 package, particularly on EN 1993-1-3. On this purpose, the book contains the essentials of theoretical background and design rules for cold-formed steel sections and sheeting, members and connections.

Online-Order: www.ernst-und-sohn.de

Ernst & Sohn
A Wiley Company

Ernst & Sohn
Verlag für Architektur und technische Wissenschaften GmbH & Co. KG

Customer Service: Wiley-VCH
Boschstraße 12
D-69469 Weinheim

Tel. +49 (0)6201 606-400
Fax +49 (0)6201 606-184
service@wiley-vch.de

*€ Prices are valid in Germany, exclusively, and subject to alterations. Prices incl. VAT. Books excl. shipping. Journals incl. shipping. 0199200006_dp

Editorial

The "Concrete Yearbook" is a very important source of information for engineers involved in design, analysis, planning and production of concrete structures. It is published on a yearly basis and offers chapters devoted to various subjects with high actuality. Any chapter gives extended information based on the latest state of the art, written by renowned experts in the areas considered. The subjects change every year and may return in later years for an updated treatment. This publication strategy guarantees, that not only the most recent knowledge is involved in the presentation of topics, but that the choice of the topics itself meets the demand of actuality as well.

For decades already the themes chosen are treated in such a way, that on the one hand the reader is informed about the backgrounds and on the other hand gets acquainted with practical experience, methods and rules to bring this knowledge into practice. For practicing engineers, this is an optimum combination. Engineering practice requires knowledge of rules and recommendations, as well as understanding of the theories or assumptions behind them, in order to find adequate solutions for the wide scope of problems of daily or special nature.

During the history of the "Concrete Yearbook" an interesting development was noted. In the early editions themes of interest were chosen on an incidental basis. Meanwhile, however, the building industry has gone through a remarkable development. Where in the past predominantly matters concerning structural safety and serviceability were in the centre of attention, nowadays an increasing awareness develops due to our responsibility with regard to society in a broader sense. This is reflected e.g. by the wish to avoid problems related to limited durability of structures. Expensive repair of structures has been, and unfortunately still is, necessary because of insufficient awareness of deterioration processes of concrete and reinforcing steel in the past. Therefore structural design should focus now on realizing structures with sufficient reliability and serviceability for a specified period of time, without substantial maintenance costs. Moreover we are confronted with a heritage of older structures that should be assessed with regard to their suitability to safely carry the often increased loads applied to them today. Here several aspects of structural engineering have to be considered in an interrelated way, like risk, functionality, serviceability, deterioration processes, strengthening techniques, monitoring, dismantlement, adaptability and recycling of structures and structural materials, and the introduction of modern high performance materials. Also the significance of sustainability is recognized. This added to the awareness that design should not focus only on individual structures and their service life, but as well on their function in a wider context, with regard to harmony with their environment, acceptance by society, the responsible use of resources, low energy consumption and economy. Moreover the construction processes should become cleaner, with less environmental nuisance and pollution.

The editors of the "Concrete Yearbook" have clearly recognized those and other trends and offer now a selection of coherent subjects which resort under a common "umbrella" of a broader societal development of high relevance. In order to be able to cope with the corresponding challenges the reader is informed about progress in technology, theoretical methods, new findings of research, new ideas on design and execution, development in production, assessment and conservation strategies. By the actual selection of topics and the way those are treated, the "Concrete Yearbook" offers a splendid opportunity to get and stay aware of the development of technical knowledge, practical experience and concepts in the field of design of concrete structures on an international level.

Prof. Dr. Ir. Dr.-Ing. h.c. *Joost Walraven*, TU Delft
Honorary president of the international concrete federation *fib*

1 Introduction

With the publication of the European technical guideline for the anchorage of post-installed metal fasteners in concrete (European Organization for Technical Approvals (EOTA) (1997)) for the first time it was possible to release European approvals for post-installed fasteners. The practical application of these approvals requires detailed design rules. At this time no European design provisions existed for fastenings and the development of generally acknowledged European design rules was not to be expected at short notice. Therefore, the design of fastenings had also to be covered in this guideline. The design method for post-installed fasteners published in Appendix C is based to a high extent on a guideline of the Deutsches Institut für Bautechnik (DIBt, German Institute of Construction Technology) from 1993 (Deutsches Institut für Bautechnik (1993)). During the past years Annex C was updated several times to the actual state of knowledge (European Organization for Technical Approvals (EOTA) (2010a)) and supplemented by the Technical Report TR 029 (European Organization for Technical Approvals (EOTA), 2010b) for the design of post-installed chemical fasteners. The current versions date from September, 2010.

The first European technical approvals for headed bolts were released in 2003. The design procedure for headed bolts was essentially based on Annex C of the above mentioned guideline and extended by applications specific to headed fasteners. This design method was a component of the approval document. These approval documents were replaced in 2011 by new versions which refer to the design provisions of CEN/TS 1992-4 as design procedure.

European technical approvals for anchor channels exist since 2011. They contain the design provisions of CEN/TS 1992-4 with slight improvements.

From the beginning the persons in charge were aware that the consideration of the design within the scope of an approval guideline could be only an interim solution, because after the European Construction Products Directive, EOTA was assigned to provide only European technical approval guidelines (ETAGs) for building products. The publication of European regulations for the design of construction products is within the responsibility of CEN. Hence, ETAG 001, Annex C should be transferred in the medium term into a European design standard. Finally in 2000 under the responsibility of CEN/TC 250 "Structural Eurocodes" this work started and was finalized in 2009. In May 2009 CEN/TS 1992-4 was accepted by the European Committee for standardization (CEN) for the tentative use as a pre-standard. The German version was published in August, 2009 by DIN (German Institute for Standardization) titled DIN SPEC 1021-4 (*Deutsches Institut für Normung (DIN)*, 2009).

The published set of rules CEN/TS 1992-4 is a European pre-standard (TS=Technical Specification, in the past named prEN). In this publication it is called CEN/TS. CEN/TS consists of the following five parts:

- CEN/TS 1992-4-1:2009: General
- CEN/TS 1992-4-2:2009: Headed Fasteners
- CEN/TS 1992-4-3:2009: Anchor Channels

Design of Fastenings for Use in Concrete – the CEN/TS 1992-4 Provisions. First edition.
Rainer Mallée, Werner Fuchs, Rolf Eligehausen.

- CEN/TS 1992-4-4:2009: Post-installed Fasteners – mechanical Systems
- CEN/TS 1992-4-5:2009: Post-installed Fasteners – chemical Systems.

Part 1 is valid for all types of fasteners. Parts 2 to 5 contain special rules for the respective fasteners. These parts shall be applied only in connection with Part 1.

Although CEN/TS 1992-4 is a pre-standard, it may be already applied for the design of fastenings, provided that their suitability was verified for the intended application by a ETA. The respective ETA must refer to CEN/TS and contain all data necessary for the calculation. The ETA can be a so-called European Technical Approval (ETA), a European harmonized product standard (hEN) or a suitable national standard or regulation. The use of the post-installed fasteners, headed bolts and anchor channels covered by CEN/TS is regulated currently only by European Technical Approvals which are called in the following ETA (European Technical Approval). Other ETAs are not available currently. They are also not in the planning stage.

In the following CEN/TS provisions are explained. Detailed descriptions of the load bearing behaviour and procedures for the calculation of fastenings with mechanical and chemical post-installed fasteners, headed bolts and anchor channels can be found in Eligehausen and Mallée (2000) as well as Eligehausen, Mallée, and Silva (2006).

2 Fields of application

CEN/TS covers the design of post-installed fastenings (fasteners) and cast *in situ* fasteners (headed fasteners and anchor channels) in concrete components. The following types of fasteners are considered:

- expansion fasteners, undercut fasteners, concrete screws, bonded fasteners, bonded expansion fasteners and bonded undercut fasteners
- headed bolts as well as anchor channels with stiff connection of anchorage element and channel.

In Figure 2.1 the different types of post-installed fasteners are shown schematically, Figures 2.2 and 2.3 show typical headed fasteners and anchor channels.

Torque-controlled post-installed expansion fasteners are subdivided into sleeve type and bolt (wedge) type expansion fasteners. Post-installed fasteners of the sleeve type (Figure 2.1a_1) consist of a screw or a threaded rod with nut, washer, distance sleeve, a part to prevent spinning of the fastener in the borehole as well as an expansion cone. Post-installed fasteners of the bolt type (Figure 2.1a_2) consist of a bolt, the end of which is formed to one or two cones and shows at the other end a thread, expansion segments nested in the conical area of the bolt, as well as of a nut and a washer. The fasteners are anchored by applying a defined torque. During torqueing a prestressing force is generated in the bolt or in the screw, the cone or the cones at the end of the fastener is pulled into the expansion sleeve or segments. These are pressed against the borehole wall. The frictional forces caused thereby, fix the fasteners in the bore hole. The load-transfer mechanism employed by expansion anchors is called 'friction'.

Displacement-controlled post-installed fasteners (Figure 2.1b) consist of an expansion sleeve and a conical expansion plug. The internally threaded steel sleeve allows to screw in a screw or a threaded rod. They are set via the expansion of the sleeve as controlled by the axial displacement of the expansion plug within the sleeve. This is achieved by driving the expansion plug into the sleeve with a setting tool and a hammer. Like torque-controlled expansion fasteners, displacement-controlled expansion fasteners transfer external tension loads into the base material via friction and, in the zone of the localised deformation to some degree via mechanical interlock.

Undercut fasteners develop a mechanical interlock between anchor and base material (working principle 'mechanical interlock'). For this a cylindrically drilled hole is modified to create a notch, or undercut, of a specific dimension at a defined location either by means of a special drilling tool or by the undercutting action of the fastener itself (self-undercutting fastener). The Figures 2.1c_1 and c_2 show two typical undercut fasteners which differ for example in the direction of the undercut: Undercut that widens towards the bottom of the borehole (Figurc 2.1c_1) or towards the concrete surface (Figure 2.1c_2). Undercut fasteners according to Figure 2.1c_1 consist of a threaded stud with a conical end, expansion sleeve, nut, and washer. Internally threaded versions (not illustrated) accept bolts or threaded rods. This type of undercut fasteners is anchored by driving the expansion sleeve onto the conical end. Then the expansion sleeve fills the undercut area

Design of Fastenings for Use in Concrete – the CEN/TS 1992-4 Provisions. First edition.
Rainer Mallée, Werner Fuchs, Rolf Eligehausen.

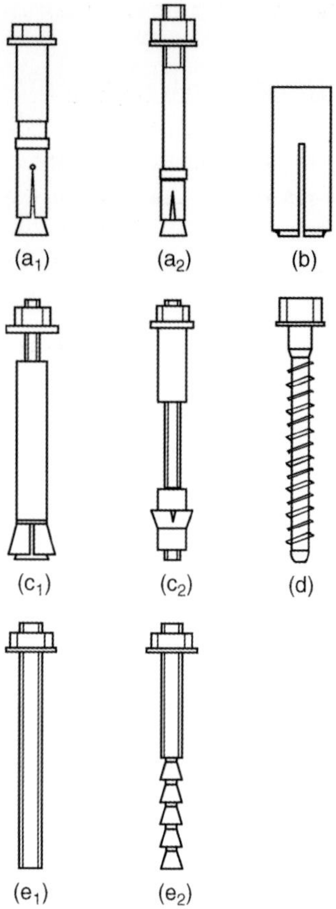

Fig. 2.1 Types of post-installed fasteners
(a_1) torque-controlled fastener (sleeve type)
(a_2) torque-controlled fastener (bolt type, also named or wedge type)
(b) deformation-controlled fastener (drop-in fastener)
(c_1) undercut fastener (undercut in the direction to the bottom of the borehole enlarged)
(c_2) undercut fastener (undercut in the direction to the concrete surface enlarged)
(d) concrete screw
(e_1) bonded fastener
(e_2) bonded expansion fastener

either produced with the help of a special tool or by cutting its undercut automatically by means of hammering or hammering/rotary action in the concrete. Undercut fasteners after Figure 2.1c_2 consist of a threaded rod with hex nut and washer, a cylindrical nut, three curved bearing segments, cone, spacer sleeve, helical spring and a plastic ring which secures the bearing segments prior to installing the anchor. After drilling a cylindrical hole, the undercut is created with the help of a special undercutting tool. Afterwards the anchor is inserted into the borehole and the bearing elements are allowed to unfold into position at the level of the undercut. Defined torqueing of the fastener brings the bearing segments into contact with the supporting surfaces.

Concrete screws (Figure 2.1d) are screwed into pre-drilled holes with the help of a special impact power screwdriver, an electric power screwdriver, a hammer drill equipped with an adapter in rotary mode or a customary torque wrench. They show typically a hardened special thread to allow the process of cutting the threads into the concrete. The diameter of the drilled hole is matched to the geometry of the screw so that the thread cuts into the concrete and an external force can be transferred to the concrete by means of mechanical interlock.

Bonded fasteners (Figure 2.1e_1) consist of a threaded rod, a hexagonal nut and a washer or an internally threaded sleeve to accept threaded parts as well as a chemical mortar as bonding material. The bonding materials may consist of polymer resins, cementitious materials, or a combination of the two. A distinction can be made between so-called capsule fasteners, in which the constituent bonding materials are contained in glass capsules or foil pouches, and injection systems. In case of capsule fasteners the mortar is mixed by driving (rotary/hammer mode of the drill) the threaded rod or internally threaded sleeve into the borehole. In case of injection systems the chemical mortar is pre-packaged in cartridges and und mixed via special mixing nozzles during injection into the borehole. The tension load is transferred to the base material by means of bond. This load-transfer mechanism is called 'chemical interlock'.

Bonded expansion fasteners (Figure 2.2e_2) employ a unique anchor rod geometry with multiple conical surfaces, nut and washer as well as an adhesive mortar as bonding agent. The mortar is delivered again in glass cartridges, foil pouches or cartridges. The anchor rod is either coated, or shows a smooth and hard surface to prevent adhesion between the anchor rod and the bonding material. After hardening of the mortar the bonded expansion anchor is pre-stressed by applying a defined installation torque. The hardened bonding material is thereby split into single mortar segments which work, in principle, like expansion sleeves of a post-installed expansion fastener. When a tensile force is applied to the anchor rod, the cones are displaced upward relative to the position of the hardened mortar and frictional forces originate between mortar sleeve and borehole wall (working principle 'friction').

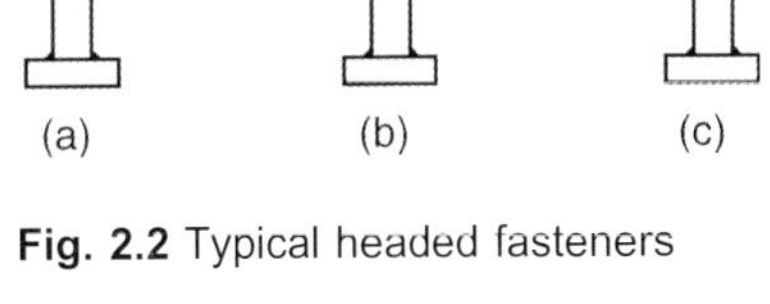

Fig. 2.2 Typical headed fasteners

Figure 2.2 shows typical headed fasteners. Headed fasteners after Figure 2.2a show a thread at the upper end for connecting with the attachment. The configurations of Figure 2.2b consist of a steel plate with headed studs butt-welded on. Headed studs are usually welded on using drawn arc stud welding. After retracting the formwork from the concrete component the attachment is welded or screwed to the embedded steel plate. Headed fasteners after Figure 2.2c show at the upper end an internal thread sleeve to accept screws or threaded rods. Headed fasteners derive their tensile resistance from mechanical interlock between the anchor head and the hardened concrete (working principle 'mechanical interlock').

Anchor channels (Figure 2.3) consist of a cold-formed or hot-rolled steel channel equipped with special anchor fittings. These channels, filled with rigid urethane foam to prevent concrete intrusion, are attached directly to the inside of the formwork. Following removal of the formwork and of the rigid foam, a variety of components can be attached with the aid of special T-headed bolts. Transfer of the load back into the concrete in case of anchor channels is generally achieved by way of T-, I-shaped or headed anchors welded or forged to the channel. However, there are also anchor channels available in which the load is transferred into the base material by way of loops of steel with tabs that are passed through the back of the channel and bent. This type of anchorage is not covered by CEN/TS since the anchor is not connected stiff to the channel and might become effective only after a certain degree of displacement of the channel. Anchor channels with serrated lips to transfer shear loads along the length of the channel are also not covered. At the time of the development of CEN/TS the state of the knowledge was not sufficient to allow for the standardisation of the design method.

CEN/TS applies to fasteners with a minimum diameter or a minimum thread size of 6 mm (M6) or a corresponding cross section and with a nominal steel tensile strength $f_{uk} \leq 1000\ \text{N/mm}^2$. In general, the minimum embedment depth should be $h_{ef} \geq 40$ mm. The actual value for a particular fastener might be taken from the relevant ETA for example a European Technical Approval (ETA).

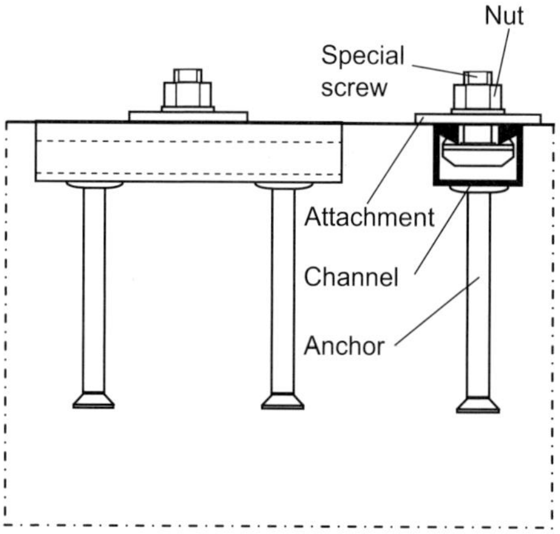

Fig. 2.3 Typical anchor channel

Furthermore CEN/TS applies to fasteners with established suitability for the specified application in concrete covered by provisions such as an ETA. CEN/TS is intended for applications in which the failure of fastenings will:

- result in collapse or partial collapse of the structure, or
- cause risk to human life, or
- lead to significant economic loss.

If the above conditions are given "safety related applications" exist and the employed fasteners must be prequalified demonstrated by an ETA and be designed accordingly. In all other cases the choice and the installation of the fasteners is performed directly by the user according to craftsmanship and knowledge, a detailed calculation of the fastening is not carried out, in general.

CEN/TS is valid for applications which fall within the scope of the series EN 1992. In applications where special considerations apply, for example nuclear power plants or civil defense structures, modifications may be necessary.

The support of the fixture may be either statically determinate or statically indeterminate, where each support may consist of one fastener or a group of fasteners. The design of the attachment for example the steel plate is not covered, requirements for stiffness and ductility of the attachment are given, however.

CEN/TS provides design methods for fasteners for structural purpose which are used to transmit actions locally into the concrete structure. They apply to single fasteners and groups of fasteners whereas it is assumed that a fastening group consists only of fasteners of the same type and size. Figure 2.4 shows possible configurations of post-installed fasteners and headed fasteners. Distinction is to be made between fastenings with and without hole clearance.

In case of post-installed fasteners no hole clearance may be considered if any inevitable gap occurring between the fastener and the fixture due to installation is filled with mortar of sufficient compression strength or eliminated by other suitable means such as snap rings.

Headed fasteners without hole clearance shall be welded to the fixture or screwed into the fixture.

If no hole clearance exists, single fastenings as well as groups with two to nine fasteners may be accomplished independent of the edge distance (Figure 2.4a). For fastenings with hole clearance and more than two fasteners in a row the distribution of a shear load on the single post-installed fasteners or headed fasteners of the group is not predictable with certainty, because their positions within the holes of the attachment can be very different by chance. Fasteners, positioned not concentric in the hole of the fixture but in direct contact with the attachment in the direction of the shear load are engaged from the beginning. On the other hand fasteners which have no initial contact with the fixture after the installation due to the hole clearance participate in the shear load transfer only if the annular gap is overcome by a certain displacement of the fixture and after a suitable deformation of the loaded post-installed fasteners. This can lead to the fact that the single fasteners of a group participate clearly differently in the transfer of

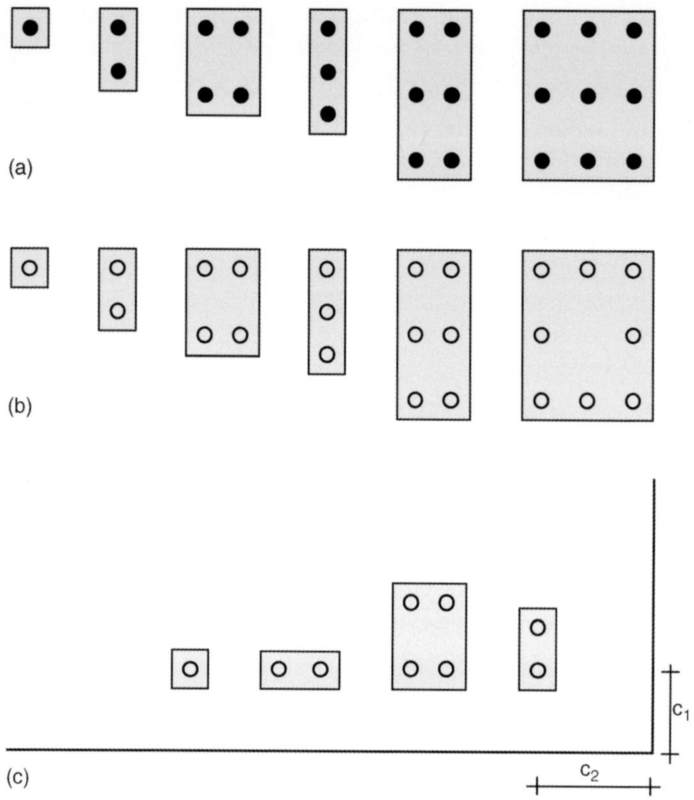

Fig. 2.4 Permissible configurations of fastenings according to CEN/TS 1992-4
(a) fastenings without hole clearance, all edge distances
(b) fastenings with hole clearance situated far from edges ($c_1 \geq 10\, h_{ef}$ and $c_1 \geq 60\, d_{nom}$, $c_2 \geq 10\, h_{ef}$ and $c_2 \geq 60\, d_{nom}$)
(c) fastenings with hole clearance situated near to an edge ($c_1 < 10\, h_{ef}$ or $c_1 < 60\, d_{nom}$, $c_2 < 10\, h_{ef}$ or $c_2 < 60\, d_{nom}$)

shear loads and in case of applications with fastenings under shear load close to the edge and the brittle failure mode concrete edge failure yield problems. Hence CEN/TS permits applications close to the edge ($c < 10 \cdot h_{ef}$ or $c < 60 \cdot d_{nom}$) (the greater of both values governs) only configurations with single fasteners or groups with two or four post-installed fasteners or headed fasteners (Figure 2.4c). Fastenings with three or six to nine post-installed fasteners or headed fasteners and hole clearance are only allowed, if the edge distance is large enough to preclude concrete edge failure. This is the case if all edge distances are $c \geq 10 \cdot h_{ef}$ and $c \geq 60 \cdot d_{nom}$ (Figure 2.4b).

This regulation limits the range of application compared with ETAG 001, Annex C, third amendment dated August, 2010 (European Organization for Technical Approvals (EOTA) (2010)). After ETAG 001 configurations according to Figure 2.4b may be carried out also with small edge distances if no shear loads act. This clear enhancement

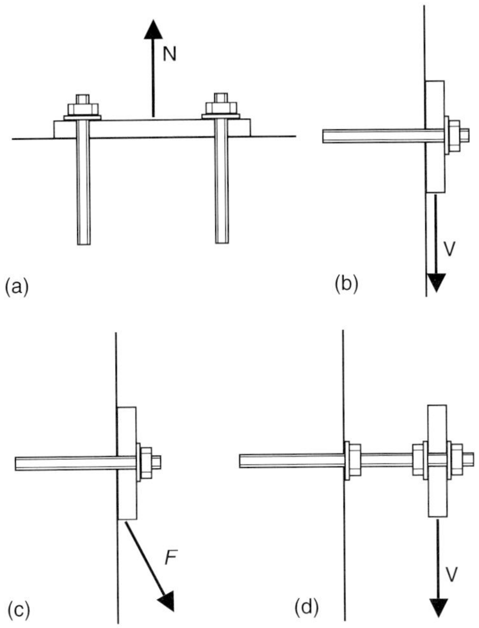

Fig. 2.5 Forces acting on fastenings
(a) tension load
(b) shear load
(c) combined tension load and shear load
(d) shear load with lever arm (stand-off installation)

of the field of application should be considered in the transfer from CEN/TS in a European Standard.

The loads acting on a fastening can be static, cyclic (causing fatigue failure) or seismic. The suitability of the fastener type to resist either cyclic or seismic loading is stated in the relevant ETA. Furthermore fasteners can be subjected to bending moments.

The loading on the fastener resulting from the actions on the fixture (e.g. tension, shear, combined tension and shear, bending moments in one or two directions or torsion moments or any combination thereof) will generally be axial tension and/or shear (Figures 2.5a to c). If a shear load with a lever arm acts (stand-off installation, Figure 2.5d), the fastener is loaded, in addition, by a bending moment.

The loads acting on the concrete component serving as anchorage ground can be static, cyclic (causing fatigue failure) or seismic. However, if the concrete member is subjected to cyclic or seismic loading only certain types of fasteners may be allowed. This is stated in the corresponding ETA (ETA).

CEN/TS is valid for concrete members using normal weight concrete with strength classes in the range C12/15 to C90/105. The range of concrete strength classes in which

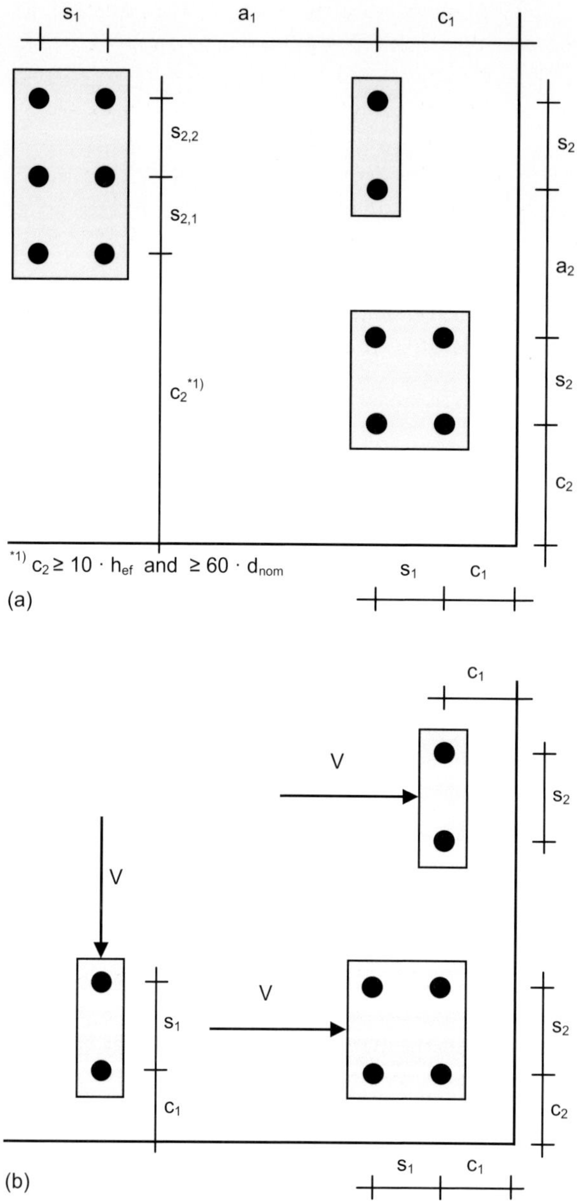

Fig. 2.6 Definition of fastener spacing and edge distance
(a) fastenings under tension load
(b) fastenings under shear loads

particular fasteners may be used is given in the relevant ETA and may be more restrictive than stated above. The strength classes for post-installed fasteners are usually in the range C20/25 to C50/60. In the region of the fastening, the concrete may be cracked or non-cracked. The condition of the concrete should be determined by the designer. In general, it is always conservative to assume that the concrete is cracked. The probability that fasteners are located in cracks is high since cracks might develop also due to restraint of intrinsic imposed deformations (e.g. shrinkage of concrete) or extrinsic imposed deformations (e.g. due to displacement of support or temperature variations).

Non-cracked concrete may be assumed if it is proven that under service conditions the fastener with its entire embedment depth is located in non-cracked concrete. This will be satisfied if Equation 2.1 is observed (compressive stresses are negative):

$$\sigma_L + \sigma_R \leq \sigma_{adm} \tag{2.1}$$

with:

σ_L stresses in the concrete induced by external loads including fastener loads.
σ_R stresses in the concrete due to restraint of intrinsic imposed deformations (e.g. shrinkage of concrete) or extrinsic imposed deformations (e.g. due to displacement of support or temperature variations). If no detailed analysis is conducted, then $\sigma_R = 3\,\text{N/mm}^2$ should be assumed.
σ_{adm} admissible tensile stress for the definition of non-cracked concrete.

The stresses σ_L and σ_R should be calculated assuming that the concrete is non-cracked. For concrete members which transmit loads in two directions (e.g. slabs, walls and shells) Equation 2.1 shall be fulfilled for both directions. The value of σ_{adm} may be found in a Country's National Annex as NDP (NDP = national determined parameter). The recommended value is $\sigma_{adm} = 0$.

The consideration of the stress σ_R in Equation 2.1 should guarantee that also by unintentional restraint of deformations the probability for the presence of cracks in the concrete is very low. The value $\sigma_R = 3\,\text{N/mm}^2$ is also used in EN 1992-1-1:2004 (Eurocode 2) for the determination of the minimum reinforcement in reinforced concrete components for the limitation of the crack width in the serviceability limit state.

For seismic design situations the concrete shall always be assumed to be cracked in the region of the fastening.

In Figure 2.6 spacing and edge distances of fasteners are defined. In case of shear loads (Figure 2.6b) it is to be observed that the indexes of spacing and edge distances depend on the direction of the shear load: Index 1 is valid for distances in the direction of and index 2 for distances perpendicular to the direction of the shear load.

3 Basis of design

3.1 General

Fastenings shall be designed according to the same principles and requirements valid for structures given in EN 1990:2002 (Eurocode 0).

For the fasteners the following limit states shall be verified:

- ultimate limit state,
- serviceability limit state.

Furthermore the durability of the fastening for the intended use shall be demonstrated.

Fasteners shall sustain all actions and influences likely to occur during execution and use (ultimate limit state). Depending on the load direction (tension and/or shear load) fasteners might fail by different failure modes. In the ultimate limit state, verifications are required for all appropriate load directions and all relevant failure modes.

In the serviceability limit state, it shall be shown that the displacements occurring under the relevant actions are not larger than the admissible displacement.

Actions shall be obtained from the relevant parts of EN 1991:2010 (Eurocode 1) or EN 1998-1:2004 (Eurocode 8) and the corresponding National Annexes. In case of seismic actions, the supplemental information to EN 1998 given in the informative Annex E of CEN/TS and described in Section 12 of this contribution shall be observed.

Fastenings shall remain fit for the use for which they are required (durability). The design working life of the fasteners shall not be less than that of the fixture. To ensure the durability of the fastening the correct choice of the fastener material and corrosion protection are of significant importance. Therefore the environmental conditions are to be taken into consideration as well as possibilities of the inspection, the maintenance and the replacement of the post-installed fasteners. Recommendations to the durability contain the informative Annex C of CEN/TS and Section 10 of this contribution.

Where applicable the fastening should have an adequate fire resistance. It should correspond to the fire resistance of the fixture.

The safety factors for resistance and durability in this CEN/TS are based on a nominal working life of at least 50 years for the fastening.

The transfer of the loads acting on the fixture to the supports of the structure shall be considered in the design of the structure taking account of the requirements of the normative Annex A (local transmission of fastener loads into the concrete member) of CEN/TS. Annex A contains rules for the verification of the component serving as anchorage material in view of the shear resistance of the concrete member and the resistance to splitting forces induced by the fasteners to be kept in addition to the requirements of Eurocode 2.

For the design and execution of fastenings covered by CEN/TS the same quality requirements are valid as for the design and execution of structures and the attachment.

Design of Fastenings for Use in Concrete – the CEN/TS 1992-4 Provisions. First edition.
Rainer Mallée, Werner Fuchs, Rolf Eligehausen.

Therefore the design of the fastening shall be performed by qualified personnel. The resistance and reliability of fastenings are significantly influenced by the manner in which the fasteners are installed.

The partial factors given in Section 3.3 are valid only when the following conditions and the conditions given in Sections 5.1 (post-installed fasteners, mechanical systems), 6.1 (post-installed fasteners, chemical systems), 7.1 (headed fasteners) and 8.1 (anchor channels) are fulfilled:

- the installation instructions and all necessary information for correct installation shall be available on site or in the precast plant at the time the installation takes place. The installation instructions for the fastener, which are normally given in the ETA (ETA) shall be followed.
- gross errors on site shall be avoided by the use of trained personnel and adequate supervision.

A project specification shall be compiled. It shall typically include the following:

- strength class of the concrete used in the design and indication as to whether the concrete is assumed to be cracked or not cracked.
- environmental exposure, assumed in design.
- a note indicating that the number, manufacturer, type and geometry of the fasteners should not be changed without reference to the original design.

The construction drawings shall include:

- location of the fasteners in the structure, including tolerances
- number, type and embedment depth of the fasteners
- spacing and edge distance of the fastenings including tolerances. Normally these should be specified with positive tolerances only
- thickness of fixture and diameter of the clearance holes in the fixture
- position of the attachment on the fixture including tolerances
- maximum thickness of an eventual intervening layer for example grout or insulation
- reference to the manufacturer's installation instructions and special installation instructions, if applicable
- note that the fasteners shall be installed ensuring not less than the specified embedment depth.

If the above conditions are complied with, no proof testing of the fasteners is necessary.

3.2 Verifications

At the ultimate limit state and the limit state of fatigue Equation 3.1 applies:

$$E_d \leq R_d \tag{3.1}$$

with:

E_d value of design action
R_d value of design resistance

At the serviceability limit Equation 3.2 applies:

$$E_d \leq C_d \tag{3.2}$$

with:

E_d design value of fastener displacement
C_d nominal value, for example limiting displacement

The forces E_d in the fasteners should be derived as recommended in EN 1990:2002 (Eurocode 0), Section 6. Forces resulting from restraint to deformation, intrinsic or extrinsic of the attached member should be taken into account in the design of fasteners. In general actions in the fixture may be calculated ignoring the displacement of the fasteners. However, the effect of the displacement of the fasteners may be significant when a statically indeterminate stiff element is fastened. This effect should be considered.

The design resistance is after Equation 3.3:

$$R_{\mathrm{d}} = R_k / \gamma_M \tag{3.3}$$

with:

R_k characteristic resistance of single fastener or group of fasteners
γ_M partial factor for resistance

The design value of fastener displacement E_d shall be evaluated from the information given in the relevant ETA (ETA). The admissible displacement C_d is to be determined by the designer. It may be assumed that it depends linearly on the applied load. In case of a combined tension and shear load the displacements resulting from tension load and shear load can be added vectorially.

3.3 Partial factors

3.3.1 General

In the following the partial factors are presented which are recommended for use in the verifications of ultimate and serviceability limit state. Partial factors of a National Annex to CEN/TS might deviate from these values. An exception forms the partial factor taking into account the installation safety of a fastener (see Section 3.3.3.1) which is established in an ETA and shall not be changed. Up to now the recommended partial factors were changed by no country.

3.3.2 Actions

Partial factors to be used are stated in EN 1990:2002 (Eurocode 0), Annex A or the corresponding National Annex. They are $\gamma_G = 1.35$ in case of dead load and $\gamma_Q = 1.50$ in case of live load. The recommended values for indirect actions are $\gamma_{ind} = 1.2$ for concrete failure and $\gamma_{ind} = 1.0$ for other modes of failure. In case of fatigue loading $\gamma_{fat} = 1.0$ is recommended.

3.3.3 Resistance

3.3.3.1 Ultimate limit state

Steel failure

The partial factors γ_{Ms} for steel in case of fasteners, anchors and special screws of anchor channels under tension and shear loading with and without lever arm can be found in the relevant ETA. They are calculated according to Equations 3.4 and 3.5:

Tension loading:

$$\gamma_{Ms} = 1.2 \cdot f_{uk}/f_{yk} \geq 1.4 \tag{3.4}$$

Shear loading with and without lever arm:

$$\gamma_{Ms} = 1.0 \cdot f_{uk}/f_{yk} \geq 1.25 \qquad f_{uk} \leq 800\,\text{N/mm}^2 \text{ and } f_{uk}/f_{yk} \leq 0.8 \tag{3.5a}$$

$$\gamma_{Ms} = 1.5 \qquad f_{uk} > 800\,\text{N/mm}^2 \text{ or } f_{uk}/f_{yk} > 0.8 \tag{3.5b}$$

Equations 3.4 and 3.5a contain the quotient from nominal tensile strength f_{uk} and nominal yield strength f_{yk}. This is necessary because the characteristic resistances for steel failure in case of tension and shear loading are based on the tensile strength f_{uk}. In the ultimatre limit state, however, the yield strength of the steel shall not be exceeded. The greater the quotient from nominal tensile strength f_{uk} and nominal yield strength f_{yk} becomes, the greater is the partial factor γ_{Ms}. That is how in the ultimate state a sufficient distance of the load-carrying capacity to the yield strength is ensured.

For anchor channels due to their design further verifications are required. The corresponding partial factors are:

Verification of the connection between anchor and channel:

$$\gamma_{Ms,ca} = 1.8 \tag{3.6}$$

Verification of the local failure of the anchor channel by bending of the lips in tension and shear:

$$\gamma_{Ms,l} = 1.8 \tag{3.7}$$

Bending of the channel:

$$\gamma_{Ms,flex} = 1.15 \tag{3.8}$$

If supplementary reinforcement is inserted to increase the load-carrying capacity of headed fasteners or anchor channels, then for the verification of the reinforcement the following partial factor shall be used:

$$\gamma_{Ms,re} = 1.15 \tag{3.9}$$

Concrete failure

The partial factor γ_{Mc} for concrete failure covers concrete cone failure, local concrete failure at the front side of the component (blow-out) under tension loading as well as concrete failure behind the fastener (pry-out) and concrete edge failure under shear load. The partial factor γ_{Msp} covers splitting failure. The values γ_{Mc} and γ_{Msp} are given in the relevant ETA. The value γ_{Mc} is usually recommended for the partial factor γ_{Msp}.

The partial factor γ_{Mc} consists of two components:

$$\gamma_{Mc} = \gamma_c \cdot \gamma_{inst} \tag{3.10}$$

The first value corresponds to the partial factor for concrete under compression. In agreement with EN 1992-1-1:2004 (Eurocode 2) $\gamma_c = 1{,}5$ is recommended. This value is also valid for fatigue and seismic loading. However, for seismic strengthening and repair of existing structures the partial factor for concrete γ_c in Equation 3.10 may be reduced according to the relevant clauses of EN 1998-1:2004 (Eurocode 8).

The second value γ_{inst} considers the installation safety of the fastener and is determined for post-installed fasteners on the basis of the results of installation safety tests after the approval guideline (European Organization for Technical Approvals (EOTA) (1997)). It is a value established for the specific product and, hence, given in the ETA of the respective product. The value for γ_{inst} shall also not be changed in the National Annex of standards.

In case of post-installed fasteners the partial factor for installation considers to which extent they react to installation inaccuracies which are not always avoidable on site. Examples are deviations of the required torque with post-installed torque-controlled expansion fasteners, deviations of the necessary displacement of the plug in case of post-installed displacement-controlled expansion fasteners or tolerances of the diameter of the tip of the used drill. The more sensitively a post-installed fastener reacts to these deviations, the greater the partial factor for installation becomes. For tension load is:

$$\gamma_{inst} = 1.0 \text{ for systems with high installation safety} \tag{3.11a}$$

$$\gamma_{inst} = 1.2 \text{ for systems with normal installation safety} \tag{3.11b}$$

$$\gamma_{inst} = 1.4 \text{ for systems with low but still acceptable installation safety} \tag{3.11c}$$

In case of shear loading it is always:

$$\gamma_{inst} = 1.0 \tag{3.12}$$

Headed fasteners and anchor channels are considered systems with high installation safety if the conditions in Section 3.1 as well as Sections 7.1 (headed fasteners) and 8.1 (anchor channels) are fulfilled. Then high installation safety $\gamma_{inst} = 1.0$ may be assumed for all load directions and no special installation safety tests are necessary.

Pull-out/pull-through failure

The partial factor γ_{Mp} for pull-out/pull-through under tension loading is given in the relevant ETA. In general for the partial factor γ_{Mp} the value for γ_{Mc} is recommended.

Limit state of fatigue

The partial factors $\gamma_{Ms,fat}$, $\gamma_{Mc,fat}$, $\gamma_{Msp,fat}$ and $\gamma_{Mp,fat}$ are given in the relevant ETA. The recommended value for steel is $\gamma_{Ms,fat} = 1.35$. The values for $\gamma_{Mc,fat} = \gamma_{Msp,fat} = \gamma_{Mp,fat}$ are calculated according to Equation 3.10.

3.3.3.2 Serviceability limit state

For the partial factor for serviceability limit state the value $\gamma_M = 1.0$ is recommended.

4 Derivation of forces acting on fasteners

4.1 General

The loads acting on a fixture (tension and compression forces, bending moments, shear loads and torsion moments) shall be transferred to the fasteners as statically equivalent tension and shear forces.

Prying forces C due to eccentricities (Figure 4.1a) or arising with deformation of the fixture and displacement of the fasteners (Figure 4.1b) should be explicitly considered in the design of the fastening. In the second case prying forces are avoided by using rigid fixtures or stiffeners welded to the steel plate.

When a bending moment and/or a compression force act on a fixture, which is in contact with concrete or mortar, a friction force develops in the interface. If a shear force is also acting on a fixture, this friction will reduce the shear force on the fastener. The sum of the shear load transferred to the concrete member via fastener and friction, however, is always equivalent to the acting load and therefore remains constant. This means, that the friction reduces the shear load acting on the fasteners, but on the other hand the forces on the concrete will not alter (e.g. for the verification of concrete edge failure). As it is difficult to quantify with confidence the effect of friction on the resistance, in CEN/TS friction forces are neglected in the design of the fastenings.

In general, elastic analysis may be used for establishing the loads on individual fasteners both at ultimate and serviceability limit states. For ultimate limit states plastic analysis for headed and post-installed fasteners may be used, if the ductility requirements for the fasteners and the conditions of CEN/TS, Annex B are observed (see Section 9).

4.2 Tension loads

4.2.1 Tension loads on fastenings with post-installed fasteners and headed fasteners

According to the theory of elasticity a linear distribution of strains across the fixture and a linear relationship between strains and stresses exists (Figure 4.2). This assumption is valid only if the fixture is rigid and does not deform significantly. The base plate should remain elastic under design actions and its deformation should be compatible with the displacement of the fasteners.

CEN/TS contains no information on the verification of a stiff base plate. Nevertheless, this verification is important because in case of a non-rigid base plate and an acting bending moment the lever arm of the internal forces is smaller than assumed in the calculation. In case of a centroid tension load and a non-rigid base plate the post-installed fasteners or headed fasteners situated close to the point of load application have to take up higher tension loads than the further remotely located fasteners.

In both cases the load-carrying capacity of the fastening can be clearly smaller than the calculated value (Fichtner (2011)). After Mallée and Riemann (1990) the steel strength under the design loads may not exceed the design value of the yield strength.

Design of Fastenings for Use in Concrete – the CEN/TS 1992-4 Provisions. First edition.
Rainer Mallée, Werner Fuchs, Rolf Eligehausen.

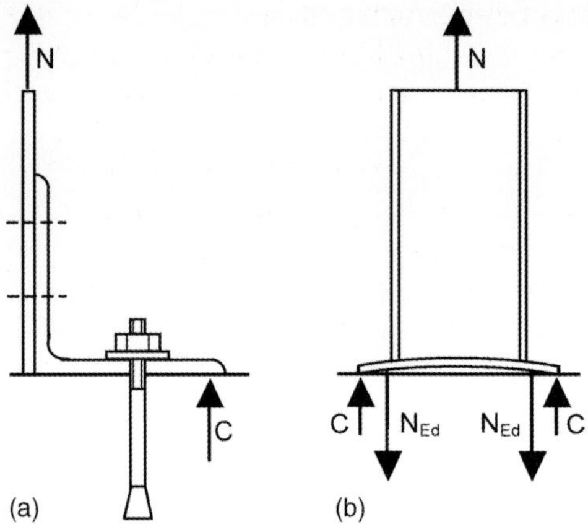

Fig. 4.1 Examples where prying forces *C* originate due to eccentricity and deformation of the attachment

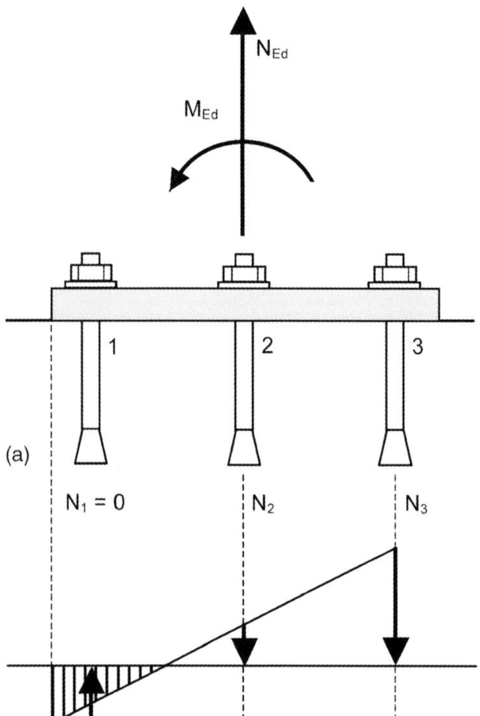

Fig. 4.2 Distribution of the forces in the fasteners and the strains below a steel plate resisting a tensile force and a bending moment (Eligehausen, Mallée, and Silva (2006))

Furthermore it shall be averaged in the points of discontinuity in the area of the corners of the built up steel section over the width $b_i = t + s$ with t the thickness of the base plate and s the wall thickness of the built up profile. A load distribution under 45° is assumed in this approach. After Mallée and Burkhardt (1999), however, the steel strength may be averaged over the width $b_i = 2 \cdot t + s$. This corresponds to a load distribution with an inclination of about 27°. By this approach it is ensured that the behaviour of the anchor plate is elastic under the design loads. This is sufficient in most applications. In unfavourable cases (e.g. small attachments in proportion to the size of the steel plate or attachments eccentrically situated on the steel plate), in addition, the deformations of the steel plate should be compared to the expected displacements of the fastener. Detailed information is to be found in Fichtner (2011).

For the determination of forces of the fasteners the following assumptions may be used:

- the axial stiffness E_s A_s of all fasteners is equal. In general A_s may be based on the nominal diameter of the fastener and $E_s = 210\,000$ N/mm^2. For threaded fasteners the stressed cross section according to ISO 898 should be taken.
- the modulus of elasticity of the concrete may be taken from EN 1992-1. As a simplification, the modulus of elasticity of concrete may be assumed as $E_c = 30\,000$ N/mm^2.
- fasteners in the zone of the fixture under compression do not take forces.

The task to transform normal forces and bending moments acting on the fixture in tension loads of the fasteners can cause trouble for the designer – provided that he has no customary design software at hand. Design software determines the tension loads iterative, that is, in case of uni-directional bending combined with normal force the dilatation of the compression zone under the steel plate and the edge strains are varied so often, until the addition of the normal forces (ΣN) and the bending moments (ΣM_x) results in zero. In case of oblique bending with or without normal force, in addition, the inclination of the neutral axis line must be varied until the addition of the moments in the second direction (ΣM_y) results also in zero. It is easily understood that such iterations can be carried out impossibly by hand. For the often appearing case of a group with two or four post-installed fasteners under tension load and uni-axial bending and usual recess of the fixture the forces in the fasteners can be determined approximately if the spacing of the fasteners is assumed as the lever arm of the internal forces below the fixture. The recess of the fixture is defined as the distance between the axis of the fastening element and the edge of the fixture. As a rule it amounts to 1.5 to 2-times the hole diameter d_f in the fixture. This approach is generally conservative since it is assumed that the resultant of the compressive stresses below the fixture is located in the axis of a fastener. However, in reality the resultant is located between fastener and edge of the fixture which results in a slightly bigger lever arm of the internal forces.

4.2.2 Tension loads on fastenings with anchor channels

The load introduction in case of anchor channels differs from the load introduction with post-installed fasteners or headed fasteners, because the stiffness of a channel is lower compared to a rigid fixture in general. The distribution of tension loads acting on the channel to the anchors may be calculated using a beam on elastic support (anchors) with a partial restraint of the channel ends as statical system. The stiffness of the elastic supports corresponds to the displacement of the anchor including the displacement of the channel lips and the concrete. The resulting anchor forces depend significantly on the assumed anchor stiffness and degree of restraint.

As a simplification for anchor channels with two anchors the loads on the anchors may be calculated assuming a simply supported beam with a span length equal to the anchor spacing. The determination of the forces acting on the anchor for anchor channels with more than two anchors is based on a proposal by Kraus (2003). Then the distribution of loads to the anchors can be calculated by means of a triangle with the apex at the position of the external load and an influence length l_i. The influence length depends on the anchor spacing and the moment of inertia of the channel. It is:

$$l_i = 13 \cdot I_y^{0.05} \cdot s^{0.5} \geq s \tag{4.1}$$

with:

I_y moment of inertia of the channel [mm^4] (see ETA)
s anchor spacing [mm]

The tension forces in each anchor due to a tension load N_{Ed} acting on the channel are calculated according to Equation 4.2. Figure 4.3 gives an example:

$$N^a_{Ed,i} = k \cdot A'_i \cdot N_{Ed} \tag{4.2}$$

with:

A'_i ordinate at the position of the anchor i of a triangle with the unit height at the position of load N_{Ed} and the base length $2 \cdot l_i$

$$k = \frac{1}{\sum_1^n A'_i}$$

N_{Ed} tension load acting on the anchor channel

If several tension loads are acting on the channel a linear superimposition of the anchor forces according to Equation 4.2 for all loads should be assumed. If the exact position of the load on the channel is not known, the most unfavourable loading position should be assumed for each failure mode. This is for the case of failure of an anchor by steel rupture or pull-out the load acting over an anchor and in case of bending failure of the channel the load acting between anchors. In case of the failure mode local flexure of the channel lip the verification shall be performed exactly at the point of introduction of the external load. The bending moment in the channel due to tension loads acting on the channel may be calculated assuming a simply supported single span beam with a span length equal to the anchor spacing. This is a simplification because it neglects the influence of partial end restraints, continuous beam action for channels with more than two anchors and catenary action after yielding of the channel. The characteristic values

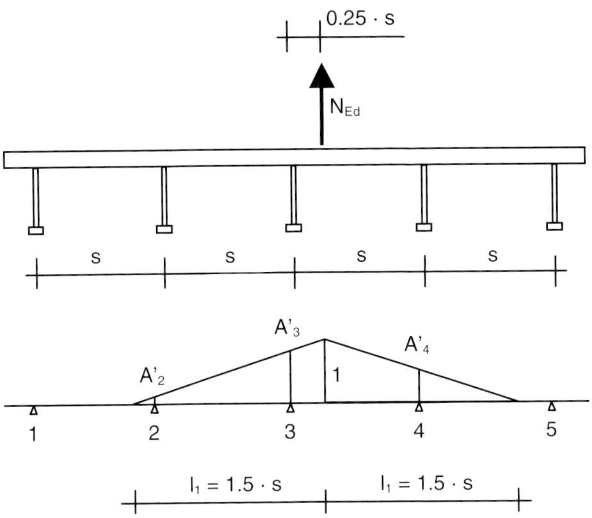

$$A'_2 = (l - 1.25 \cdot s) / l = 1/6$$
$$A'_3 = (l - 0.25 \cdot s) / l = 5/6$$
$$A'_4 = (l - 0.75 \cdot s) / l = 1/2$$
$$k = 1 / (A'_2 + A'_3 + A'_4) = 2/3$$
$$N^a_{ED,1} = N^a_{Ed,5} = 0$$
$$N^a_{Ed,2} = 2/3 \cdot 1/6 \cdot N_{Ed} = 1/9 \cdot N_{Ed}$$
$$N^a_{Ed,3} = 2/3 \cdot 5/6 \cdot N_{Ed} = 5/9 \cdot N_{Ed}$$
$$N^a_{Ed,4} = 2/3 \cdot 1/2 \cdot N_{Ed} = 1/3 \cdot N_{Ed}$$

Fig. 4.3 Example for the calculation of the anchor forces for an anchor channel with 5 anchors and an assumed influence length $l_i = 1{,}5\,s$ (Kraus (2003))

of the moments of the resistance given in the ETA take these effects into account. They may be larger than the plastic moment, calculated with the dimensions of the channel and nominal yield strength of the steel.

4.3 Shear loads

4.3.1 Shear loads on fastenings with post-installed and headed fasteners

According to CEN/TS post-installed fasteners and headed fasteners resist shear loads only if the diameter d_f in the hole of the fixture is not larger than the value given in Table 4.1. The diameter d_f is defined in Figure 4.4. Fasteners positioned in slotted holes in the direction of the shear load or in holes with a diameter d_f exceeding the

Table 4.1 Diameter of clearance hole d_f in the fixture

external diameter $d^{a)}$ or $d_{nom}^{b)}$ [mm]	6	8	10	12	14	16	18	20	22	24	27	30
diameter d_f of the clearance hole [mm]	7	9	12	14	16	18	20	22	24	26	30	33

a) bolt bears against the fixture (Figure 4.4a)
b) sleeve bears against the fixture (Figure 4.4b)

values of Table 4.1 are not considered to take up shear loads. This is due to the fact that post-installed fasteners and headed fasteners in groups with "normal" hole clearance in case of the failure modes steel failure and pry-out failure even if dissimilarly positioned in the holes (e.g. with or without contact to the fixture) deform sufficiently at ultimate state to resist uniformly to the external load. CEN/TS indirectly ensures this deformation capacity by simply limiting the tensile steel strength of the fasteners to $f_{uk} \leq 1000\,\text{N/mm}^2$. Examples of the distribution of the shear load to all post-installed fasteners or headed fasteners within a group shows Figure 4.5. In case of a too large clearance hole in the fixture it is possible that the deformation capacity of the fasteners is not sufficient to bridge the large hole before individual fasteners start to fail. Then at ultimate limit state even in case of steel failure and pry-out failure not all post-installed fasteners or headed fasteners take up shear loads and a zipper failure might occur.

Fastenings close to the edge subjected by a shear load perpendicular to the edge might fail by concrete edge break-out. Then the displacements are relatively small in the state of failure, because the concrete edge fails brittle. Therefore, it is not sure whether the deformations of the fasteners are sufficient to guarantee a load transfer to all post-installed fasteners in case of a present hole clearance meeting the requirements of Table 4.1 for the diameter d_f, before concrete edge rupture occurs. Therefore it is assumed that in a group of fasteners only the fasteners close to the edge resist shear loads. This corresponds to the case that after installation the post-installed fasteners close to the edge are in contact with the fixture. The remotely located post-installed fasteners of the group are without contact. Figure 4.6 shows such a case at the example of a group with two post-installed fasteners. Examples of the distribution of the shear load for the verification of concrete edge failure shows Figure 4.7.

In case of groups with post-installed fasteners without hole clearance close to the edge the approach that only the fasteners close to the edge are effective can be conservative for the verification of concrete edge failure. Then it is to be expected that with increasing shear force first one crack originates from the post-installed fasteners close to edge which does not cause, however, to the failure of the whole group. Only if with increasing shear force another crack develops beginning from the remote post-installed fasteners, the group fails. However, this model is valid only for groups subjected by pure shear loads. If a tension force acts on the anchor plate or fixture simultaneously, then the crack developing from the post-installed fasteners close to edge can already lead to failure, provided that no suitable edge reinforcement exists which keeps the

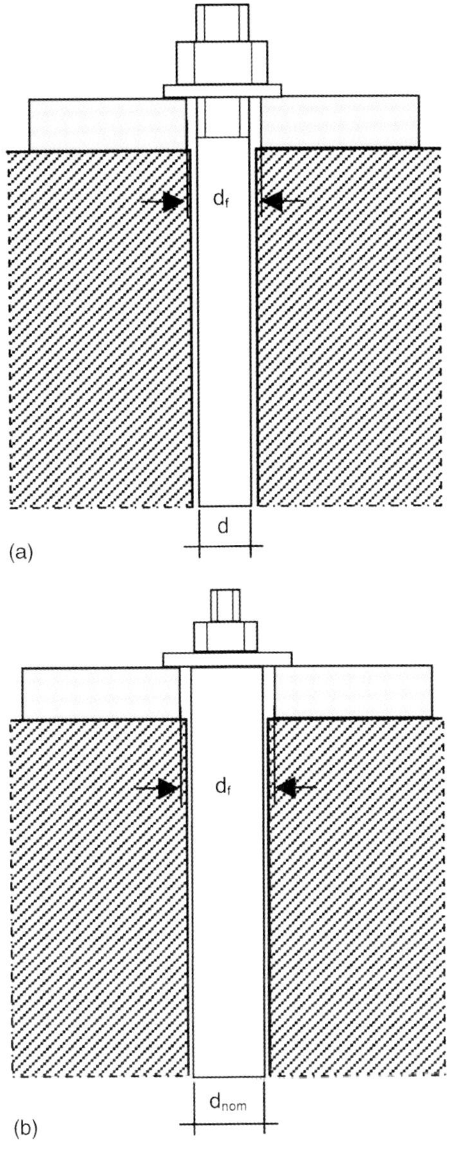

Fig. 4.4 Definition of the hole d_f for fastenings with hole clearance
(a) bolt is assumed to bear against the fixture (bolt type fastener)
(b) sleeve is assumed to bear against the fixture (sleeve type fastener)

crack width small or if fasteners are used which are not suitable for applications in cracked concrete. In both cases the crack close to the edge causes pull-out failure of the post-installed fasteners close to edge. Then the remote post-installed fasteners are not able to take up the tension load acting on the group, in general (Figure 4.8).

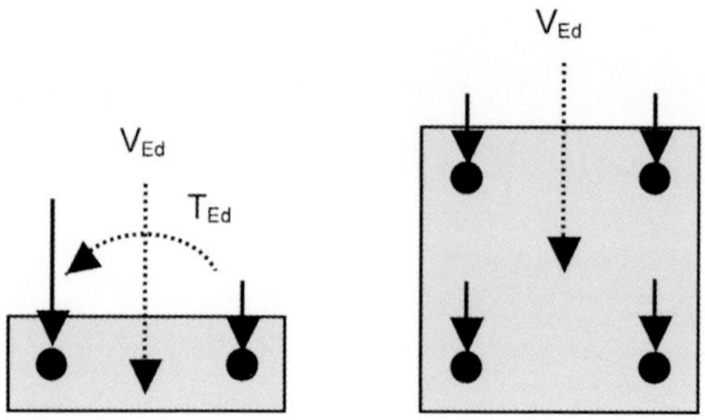

Fig. 4.5 Distribution of the shear load if all fasteners of a group resist shear loads (failure modes: Steel failure and concrete pry-out

Until now there are no design rules for the edge reinforcement to limit the width of the crack close to edge. In particular there is a lack of information on cross section, direction and anchorage of this reinforcement. Even in the literature, no test results are published. Here further research is required. As long as the open questions are not answered, the consideration of the remote post-installed fasteners in the design can only be recommended for the verification of concrete edge failure after detailed analysis by the designer.

Fastenings located close to the edge and loaded by a shear load parallel to the edge may also fail by concrete edge break-out. Indeed, the displacements are larger in the failure state compared to a shear load directed perpendicular to the edge. Hence, a shear load acting parallel to the edge may be also distributed to all post-installed fasteners and headed fasteners in fastenings without or with hole clearance (d_f according to Table 4.1).

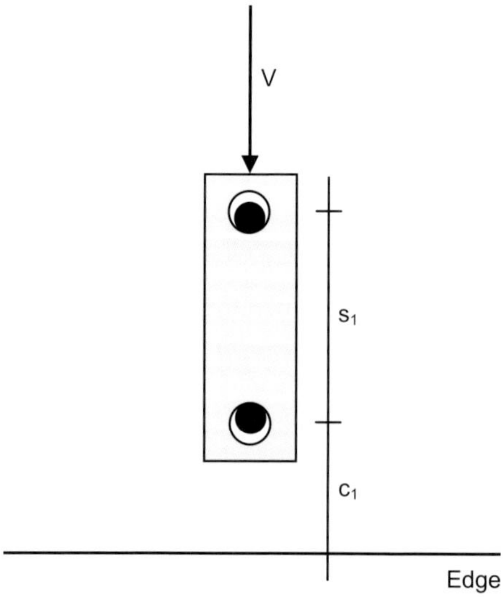

Fig. 4.6 Unfavourable positioning of the fasteners in the holes of a double fastening resisting a shear load acting perpendicular to the component edge

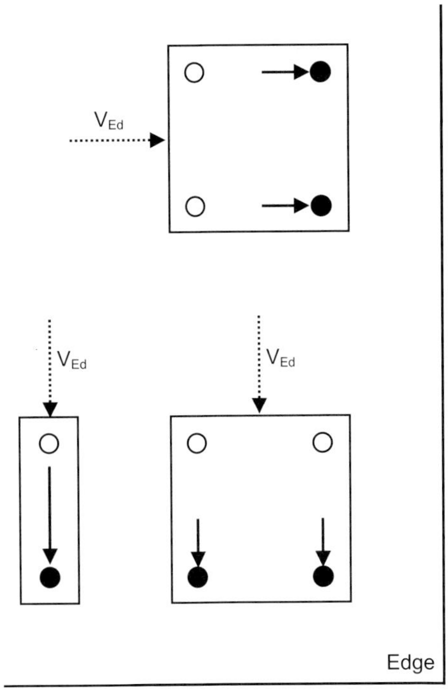

Fig. 4.7 Distribution of the shear loads when only the unfavourable fasteners close to the edge are effective (failure mode: Concrete edge failure)

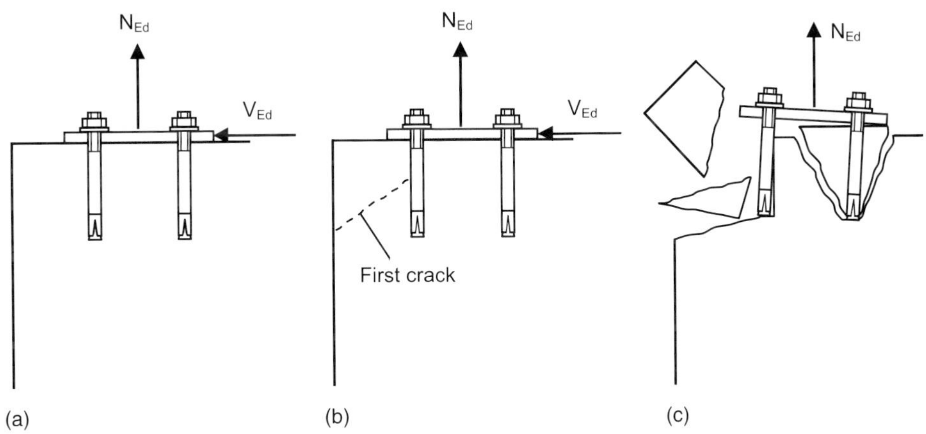

Fig. 4.8 Example of a possible failure mechanism of a fastener group close to the edge under combined tension and shear load if only the back fasteners are assumed effective for the verification of concrete edge break-out (after fib (2011))
(a) double fastening, resisting a tension and a shear force
(b) crack, originating from the fastener close to the edge as a result of the shear load
(c) tension failure of the fastener close to the edge

CEN/TS contains no information on the distribution of a torsion moment to the fasteners of a group located close to the edge and how the resulting shear forces on the fasteners are to be dealt with. Analogously to the consideration of shear forces acting on fastenings close to the edge the following approach is possible: All post-installed fasteners or headed fasteners resist the shear load independent if there is a hole clearance present (d_f after Table 4.1) or not. The shear loads acting on the single fasteners of the group are disassembled in components vertical and parallel to the edge. Then these components are treated like shear loads in general, that is the components of shear loads acting vertically to the edge are distributed to the fasteners close to the edge and the edge-parallel components are resisted by all fasteners. Figure 4.10c explains this approach by means of a quadruple fastening close to the edge subjected by a torsion moment. The authors are of the opinion that the described approach is conservative.

The design value of the shear forces of the individual fasteners of a group resulting from shear forces and torsion moments acting on the fixture may be calculated using the theory of elasticity. This is possible, because according to CEN/TS, Part 1, Section 1.2.3, in a group only fasteners of the same type and size may be used. Hence, the stiffness of all fasteners is equal. As already explained, the distribution of the shear loads on the single fasteners of a group depends on the respective failure mode. This is also to be observed in the determination of the value of the shear loads. Figure 4.9 shows examples of the failure modes steel failure and concrete pry out. In Figure 4.10 examples of the failure mode concrete edge break-out are shown. In both figures the shear force acts in the centre of gravity of the fasteners.

If the fixture is not subjected to a torsion moment, the shear loads working on the single fasteners are calculated in the examples after Figure 4.9a–c by division of the acting shear force by the number of the post-installed fasteners or headed fasteners of the group. In case of torsion (Figure 4.9d) the shear load is determined from the torsion moment, the polar inertia moment and the distance of the respective fastening element to the centre of gravity of all fasteners.

As already mentioned, for the verification of concrete edge failure only the fasteners close to the edge take up shear loads (Figure 4.7). However, this is valid only for shear forces which are directed to the free edge of the component. On the other hand a shear load acting parallel to the edge is distributed to all fasteners (Figure 4.10a). Due to the above reasons the edge-parallel component of a shear force inclined to the edge is distributed to all fasteners, while the component acting vertical to the edge is assigned only to the fasteners close to the edge (Figure 4.10b and c).

It could be proven by tests that the shear loads which are directed away from the edge cannot influence the concrete edge failure, in general, and therefore can be neglected (Mallée (2002)). Solely in special cases (small ratio of spacing to edge distance and high ratio of the characteristic resistance for concrete pry-out failure and concrete edge failure) the load causing concrete edge break-out can be also influenced unfavorably by a shear load directed away from the edge (Grosser (2012)). However, in practice these special cases appear not frequently. Therefore, for the verification of concrete edge failure the components of the shear load which are directed away from

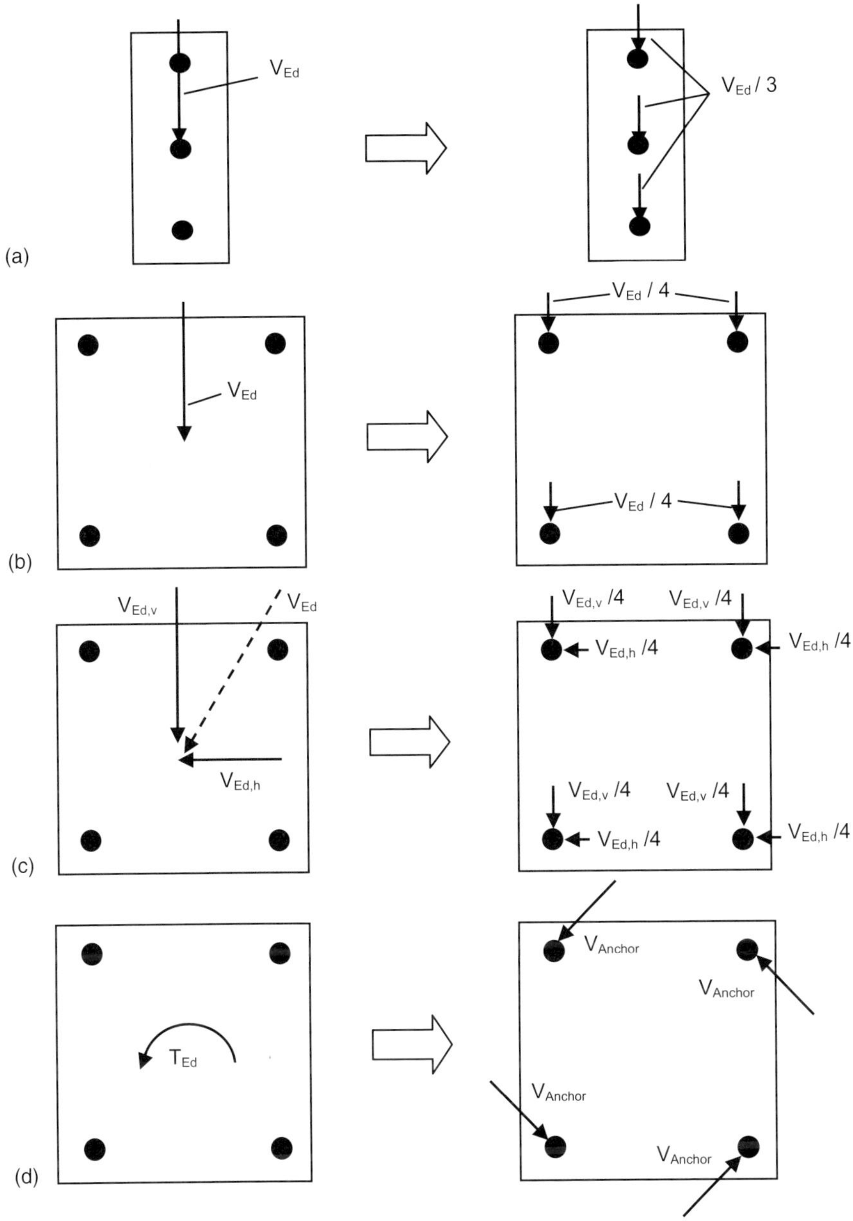

Fig. 4.9 Distribution of the shear load when all anchors of a fastening are effective (failure modes: Steel failure and concrete pry-out

(a) group with three fasteners in a row loaded by a shear force

(b) quadruple fastening, loaded by a shear force

(c) quadruple fastening, loaded by an inclined shear force

(d) quadruple fastening, loaded by a torsion moment

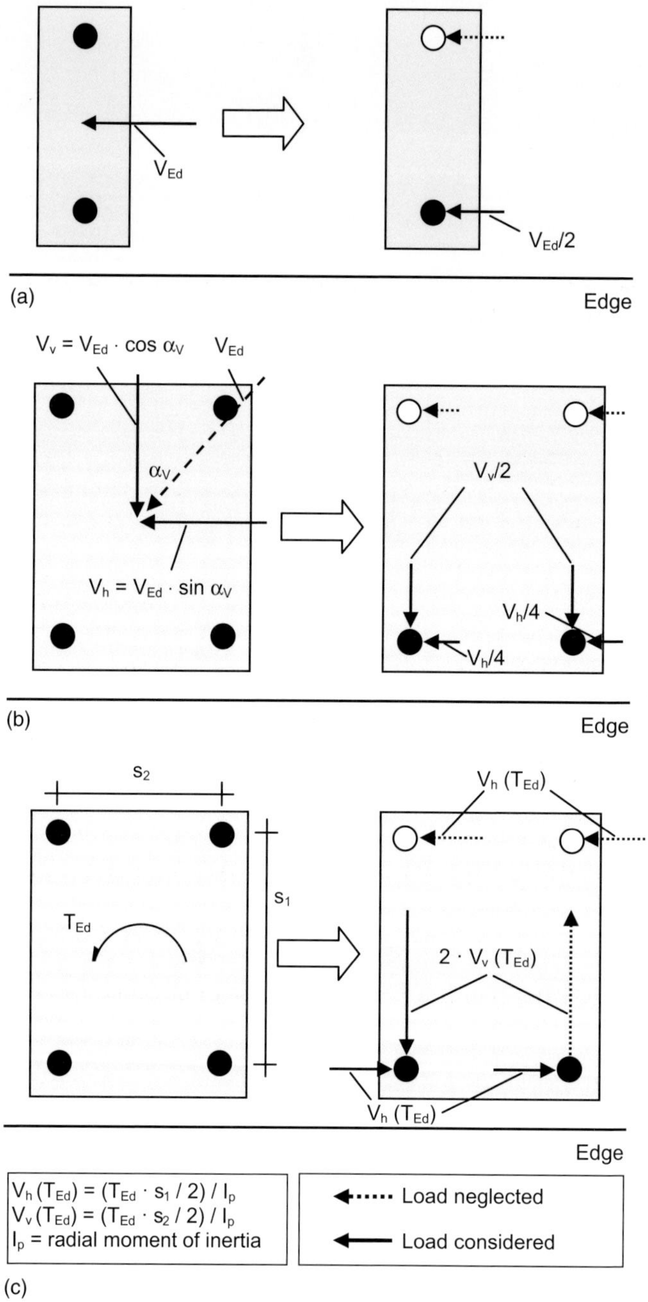

Fig. 4.10 Distribution of the shear load if only the unfavourable fasteners of a group are effective (failure mode: Concrete edge failure)

(a) double fastening, loaded by an edge-parallel shear force

(b) quadruple fastening, loaded by an inclined shear force

(c) quadruple fastening, loaded by a torsion moment

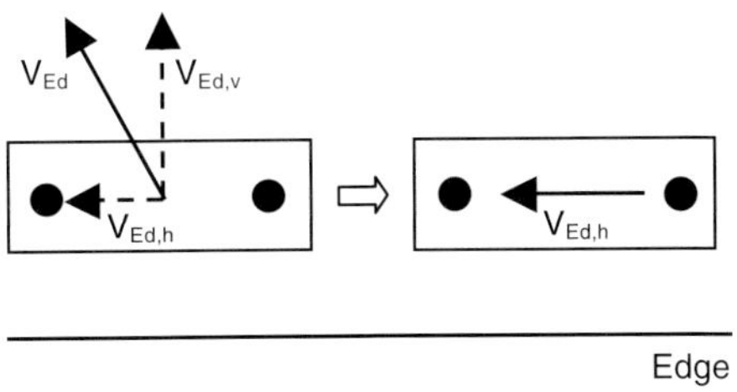

Fig. 4.11 Distribution of the shear force in the verification for concrete edge failure in case of an inclined shear load directed away from the edge

the edge are neglected in CEN/TS. Figure 4.11 shows an example. The acting shear load V_{Ed} is decomposed into components directed vertically ($V_{Ed,v}$) and parallel ($V_{Ed,h}$) to the edge. The vertical component, directed away from the edge component is not considered in the design, and the verification of concrete edge failure is only performed for the horizontal component $V_{Ed,h}$.

In the above considerations it was assumed that the baseplate or fixture is laid directly or with a thin layer of levelling grout on the concrete surface. If a thicker grout layer (Figure 4.12) is between concrete and base plate or if the fastening is mounted in stand-off installation (Figure 2.5d), then the fasteners are additionally loaded by a bending moment. This is to be considered in the verification of steel failure. According to CEN/TS shear loads acting on fastenings may be assumed to act without a lever arm if all of the following conditions are fulfilled:

a) The fixture must be made of metal and in the area of the fastening be fixed directly to the concrete without an intermediate layer or with a levelling layer of mortar with a compressive strength $\geq 30\,N/mm^2$ and a thickness $\leq d/2$.
b) The fixture is in contact with the fastener over a length of at least $0.5 \cdot t_{fix}$ (Figure 4.13).
c) The diameter d_f of the hole in the fixture is not greater than the value given in Table 4.1.

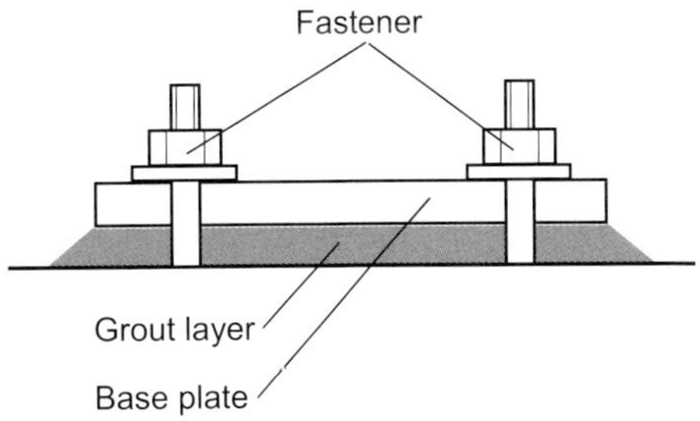

Fig. 4.12 Fixture with thick grout layer

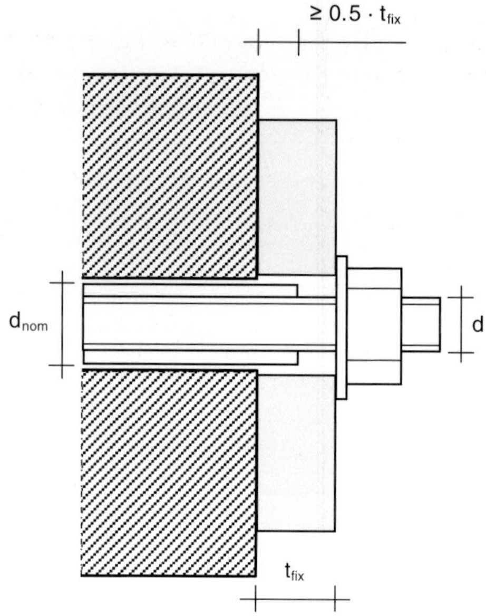

Fig. 4.13 Required bearing area of a fastener under shear loading

If the above conditions (a) to (c) are not fulfilled, the verification shall be done for a shear load acting with a lever arm (Figure 4.14):

$$l = a_3 + e_1 \tag{4.3}$$

with:

a_3 $= 0.5\ d$ (Figure 4.14a)
$= 0$ if a washer and a nut are directly clamped to the concrete surface, or if a levelling grout layer with a compressive strength $\geq 30\,\text{N/mm}^2$ and a thickness $t_{grout} > d/2$, is present
d diameter of the bolt or thread diameter
e_1 distance between shear load and concrete surface

The value a_3 considers that during the bore drilling process concrete spalling originates on the concrete surface which increases the lever arm of the shear load. The concrete spalling does not have to be taken into account if the fastener is clamped to the concrete surface by a nut and a washer or if the gap is filled with grout. If in case of cast-in headed fasteners the fixture is installed according to Figure 2.2a or c with a distance to the concrete surface, then $a_3 = 0$ may be assumed, because concrete spalling does not occur.

The design moment acting on the fastening is calculated according to Equation 4.4:

$$M_{Ed} = V_{Ed} \cdot \frac{l}{\alpha_M} \tag{4.4}$$

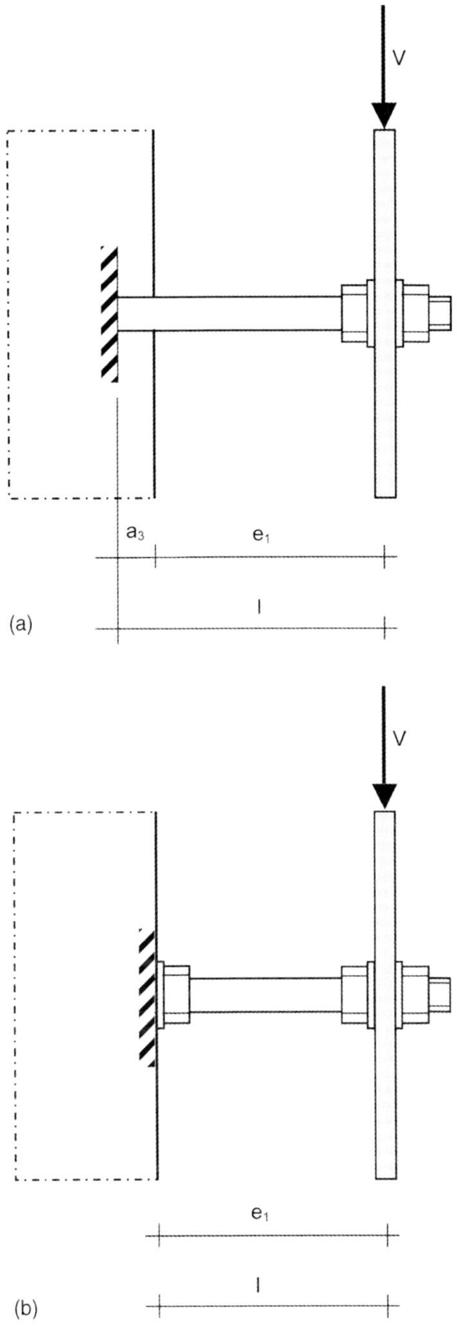

Fig. 4.14 Definition of the lever arm
(a) without restraint of the fastener at the concrete surface
(b) with restraint of the fastener at the concrete surface

with:

V_{Ed} design value of the acting shear load
l lever arm according to Figure 4.14
α_M factor to consider the restraint in the fixture

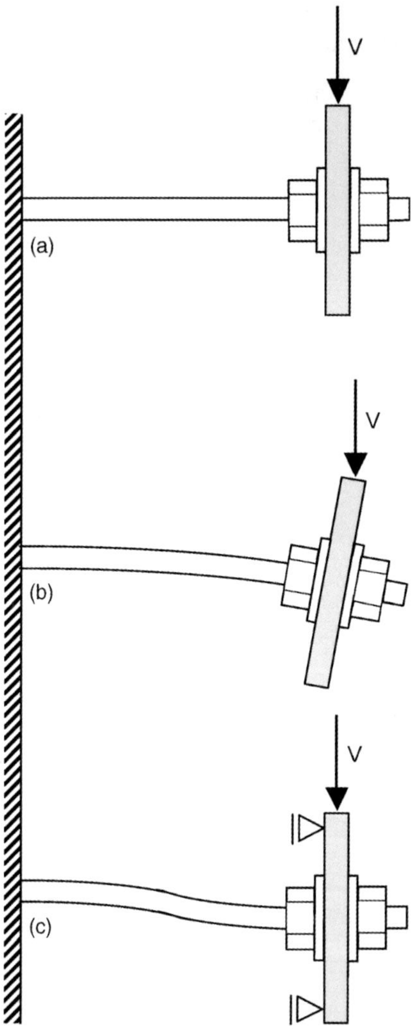

Fig. 4.15 Fasteners without and with restraint in the fixture (Comité Euro-international du Béton (1995))
(a) undeformed system
(b) deformed system without restraint ($\alpha_M = 1.0$)
(c) deformed system with full restraint ($\alpha_M = 2.0$)

α_M takes into account the degree of restraint of the fastener in the fixture (Figure 4.15). The degree of restraint should be determined by the designer according to good engineering practice. No restraint ($\alpha_M = 1,0$) should be assumed if the fixture can rotate freely (Figure 4.15b). Full restraint ($\alpha_M = 2,0$) may be assumed only if the fixture can not rotate and the fixture is clamped to the fastening by a nut and washer (Figure 4.15c). It is to be observed that the fastened element must be able to take up the restraint moment. In case of doubt it is recommended to assume $\alpha_M = 1.0$.

4.3.2 Shear loads on fastenings with anchor channels

CEN/TS covers only shear loads acting on the channel perpendicular to its longitudinal axis. They are transferred into the concrete by the channel and the anchors (Figure 4.16a). The share of shear load which is transferred by the channel depends on numerous parameters and can vary considerably (Potthoff (2008)). For reasons of simplicity especially to allow for a simple interaction between tension and shear forces acting on the channel it is assumed that the shear forces are transferred into the concrete only by the anchors located within the influence length of the load (Figure 4.16b). Therefore the shear forces of each anchor due to a shear load acting on the channel perpendicular to its longitudinal axis may be calculated as described in Section 4.2.2. The anchors resist also the tension load caused by the eccentricity between the acting shear load and the resultant of the stresses in the concrete (Figure 4.16). They thereby hinder the breaking of the channel off the concrete. This tension force is neglected in the design.

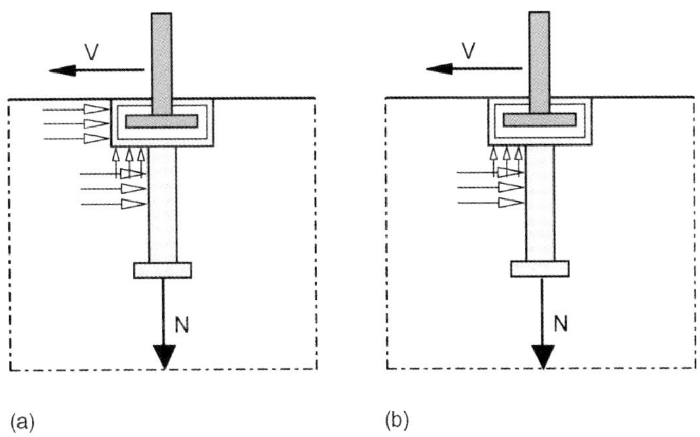

Fig. 4.16 Transfer of shear loads at anchor channels via pressures between the web of the anchor channel situated in load direction and the concrete as well as by means of the anchors (Eligehausen *et al.* (2007))
(a) shear loads are assigned to the channel and the anchor
(b) shear loads are assigned only to the anchor

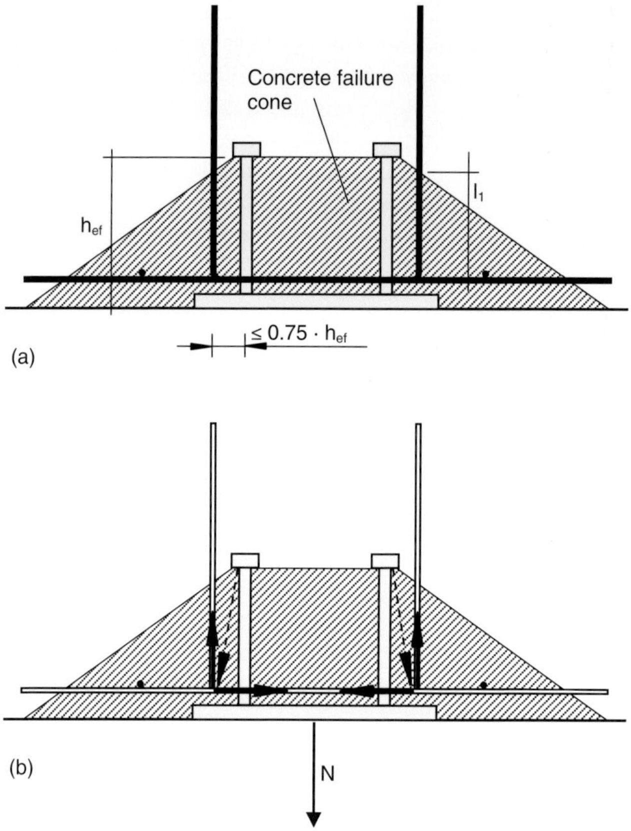

Fig. 4.17 Example of a fastener group under tension load with supplementary reinforcement to take up tension loads and strut and tie model
(a) fastening with headed fasteners and hanger reinforcement
(b) strut and tie model

4.4 Tension forces in a supplementary reinforcement

Supplementary reinforcement to increase the load bearing capacity makes sense only in case of fastenings with headed fasteners and anchor channels, because then the reinforcement can be inserted together with the fasteners before placing of concrete.

The design tension forces in the supplementary reinforcement should be established using an appropriate strut and tie model. Examples are given in Figure 4.17 for fastenings under tension loading and Figure 4.18 for fastenings with headed fasteners under shear loading.

The supplementary reinforcement for shear-loaded fastenings should be detailed in form of stirrups or loops with a mandrel diameter according to EN 1992-1-1. The supplementary reinforcement should be placed as close to the bottom of the baseplate

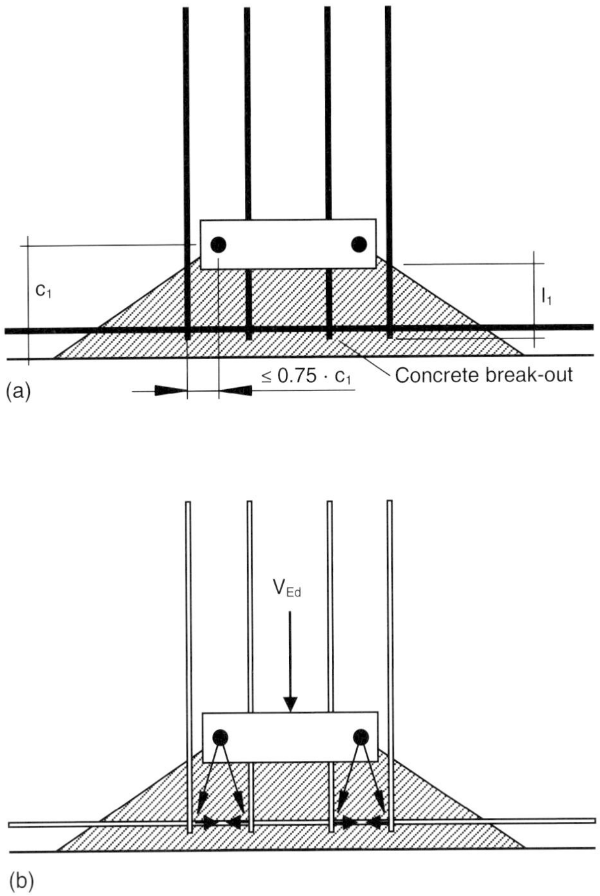

Fig. 4.18 Example of a fastener group under shear load with supplementary reinforcement to take up tension loads resulting from the shear forces and strut and tie model
(a) fastening with headed fasteners and hanger reinforcement
(b) strut and tie model.

observing the required minimum concrete cover and preferably enclose the fastener with direct contact (Figure 4.19). This yields an optimum use of the load-carrying capacity of the supplementary reinforcement. This kind of supplementary reinforcement is less effective with anchor channels because a large share of the shear force is transferred directly in the concrete by the channel. Hence, a supplementary reinforcement in the form of stirrups and loops which enclose the anchors directly is not regulated in CEN/TS, Part 3.

The supplementary reinforcement should be designed to resist the entire load acting on the fastening.

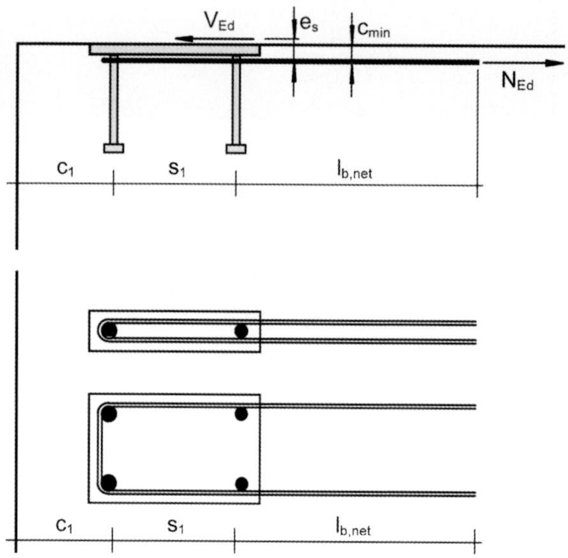

Fig. 4.19 Examples of detailing of the supplementary reinforcement in form of loops in case of shear loaded fastenings with headed fasteners

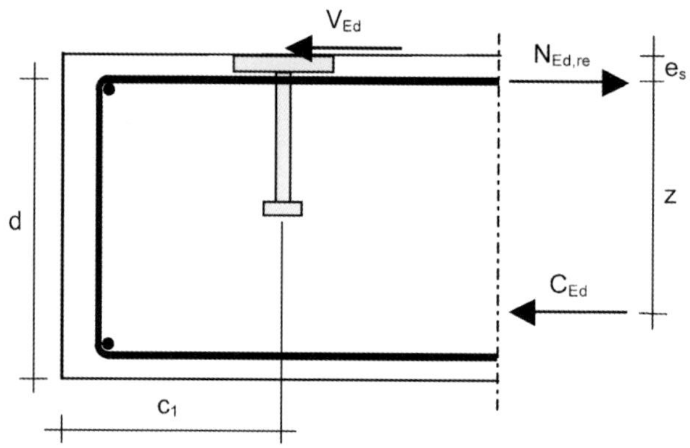

Fig. 4.20 Determination of the design tension load $N_{Ed,\ Re}$ of the supplementary reinforcement for the transfer of shear loads acting on the fixture

The eccentricity of the load transfer shall be observed in the calculation of the design tension load $N_{Ed,re}$ of the supplementary reinforcement caused by the design shear load V_{Ed} acting on the fastening (Figure 4.20) by Equation 4.5:

$$N_{Ed,re} = V_{Ed} \cdot \left(\frac{e_s}{z} + 1 \right) \tag{4.5}$$

with:

e_s distance between reinforcement and shear force acting on a fixture
z internal lever arm of the concrete member $\approx 0.85d$
d according to Figure 4.20

$$\leq min \begin{cases} 2\,h_{ef} \\ 2\,c_1 \end{cases}$$

In case of different shear forces on the fasteners of a fixture, Equation 4.5 should be solved for the shear load V^h_{Ed} of the most loaded fastener resulting in the design tension laod of the supplementary reinforcement $N^h_{Ed,re}$.

5 Verification of ultimate limit state by elastic analysis for post-installed fasteners (mechanical systems)

5.1 General

The design provisions of CEN/TS, Part 4, described in the following are valid for post-installed fasteners as shown in Figure 2.1, with exception of bonded fasteners after Figure 2.1e_1. This type of fasteners transfers the loads – in contrast to the remaining fasteners of Figure 2.1 – continuously along the entire embedment depth in the concrete. This anchorage principle requires special design rules which are described in Section 6.

For the design of post-installed fasteners (mechanical systems) in the ultimate limit state, there are three different design methods available in CEN/TS. The methods differ significantly in the degree of simplification at the expense of conservatism. In method *A* the resistance is established for all load directions and all modes of failure, using actual values of edge distance *c* to the fasteners and spacing *s* between fasteners in a group which might be smaller than characteristic values. Characteristic spacing and edge distances are the distances for ensuring the maximum characteristic resistance of a single fastener. Method *A* is most complex, however, offers almost for any application possibilities and leads to the best results. In method *B one* single value of resistance is used for *all* load directions and modes of failure. This resistance is related to the characteristic spacing and edge distances and it is permitted to use smaller values for c and s than these but the resistance should then be modified as indicated – as for method *A*. Method *B* is less complex than method *A*, however, yields more conservative results. The most simple and most conservative design procedure is method *C*. It corresponds, in principle, to method *B*, however, spacing and edge distances below the characteristic values is not permitted. Method *C* is hardly applied in practice due to its uneconomical results. Though method *B* offers less uneconomical solutions than method *C*, nevertheless, compared with method *A*, method *B* owns in particular in the design of shear-loaded fastenings considerable disadvantages. Therefore, method *B* is infrequently used in practice. Hence, in the following sections only design procedure *A* is presented.

For post-installed chemical systems, CEN/TS, Part 5 provides only one design method. Its philosophy corresponds to method *A* for post-installed mechanical systems. The same is valid for headed fasteners (CEN/TS, Part 2).

The distance between the external post-installed fasteners of neighbouring post-installed fastener groups or the distance between single fasteners or the external post-installed fasteners of groups and single fasteners shall be $a > s_{cr,N}$ ($s_{cr,N}$ = characteristic spacing). This corresponds to the necessary minimum distance, to prevent that the load bearing behaviour of adjacent post-installed fastener groups or single post-installed fasteners does not influence mutually (e.g. in case of conical concrete break-out).

The following assumptions in respect to installation have been made in CEN/TS. Prior to drilling of the borehole via visual check it shall be ensured that the concrete has been compacted adequately in the area of the fastening. The holes are drilled perpendicular

Design of Fastenings for Use in Concrete – the CEN/TS 1992-4 Provisions. First edition.
Rainer Mallée, Werner Fuchs, Rolf Eligehausen.

to the surface of the concrete by the drilling method specified by manufacturer's instructions since the drilling method can essentially influence the load-bearing behaviour of post-installed fasteners. Deviations from the drilling directions are only admissible if specifically required by the manufacturer's instructions. When hard metal hammer-drill bits are used, they should comply with ISO or National Standards. When diamond core drilling is permitted, the diameter of the segments should comply with the diameter prescribed in the manufacturer's installation instructions. Reinforcement shall not be damaged during drilling. In prestressed concrete structures it shall be ensured that the distance between the drilling hole and the prestressed reinforcement is at least 50 mm. Boreholes are to be cleaned according to the manufacturer's installation instructions. Aborted drill holes which are filled with high strength non-shrinkage mortar do not have to be considered in the design of the fasteners. This assumption is acceptable because the filled borehole for example in case of a shear load directed to the aborted hole will have similar characteristics as the intact adjacent concrete.

As a matter of principle inspection and approval of the correct installation of the fasteners is carried out by appropriately qualified personnel.

5.2 Tension load

5.2.1 Required verifications

Post-installed fasteners fail under tension load by steel rupture, pull-out/pull-through, originate a conical concrete break-out or split the concrete component (Figure 5.1). Steel failure (Figure 5.1a) is characterized by fracture in the shaft or the thread area or by rupture the sleeve. Steel failure constitutes to the greatest possible resistance of a fastener.

Pull-out failure (Figure 5.1b_1) is a failure mode where the complete fastener is pulled out of the borehole. The concrete close to the surface might be also damaged. The influence on the resistance of this secondary damage can be neglected. Pull-out failure can be expected for torque-controlled expansion fasteners in which the expansion force is too small to anchor the fastener in the borehole or for fasteners with insufficient follow-up expansion. Furthermore pull-out failure can occur if the expansion force of displacement controlled fasteners is too small. Pull-through failure (Figure 5.1b_2) appears only in case of properly functioning torque-controlled expansion fasteners. Then the expansion cone is pulled through the expansion sleeve or expansion segments. This behaviour complies with the working principle of this type of fasteners. Due to reasons of simplification CEN/TS does not distinguish between the failure modes pull-out and pull-through. The common term used for both failure modes is pull-out.

In case of conical shaped concrete break-out the fastener separates a concrete cone from the base material (Figure 5.1c_1). If, however, fasteners are commonly loaded by a steel plate and grouped with small spacing the individual failure cones overlap and a common concrete break-out cone develops (Figure 5.1c_2). If a fastener is positioned close to the edge of a concrete component, then a complete concrete cone cannot form (Figure 5.1c_3).

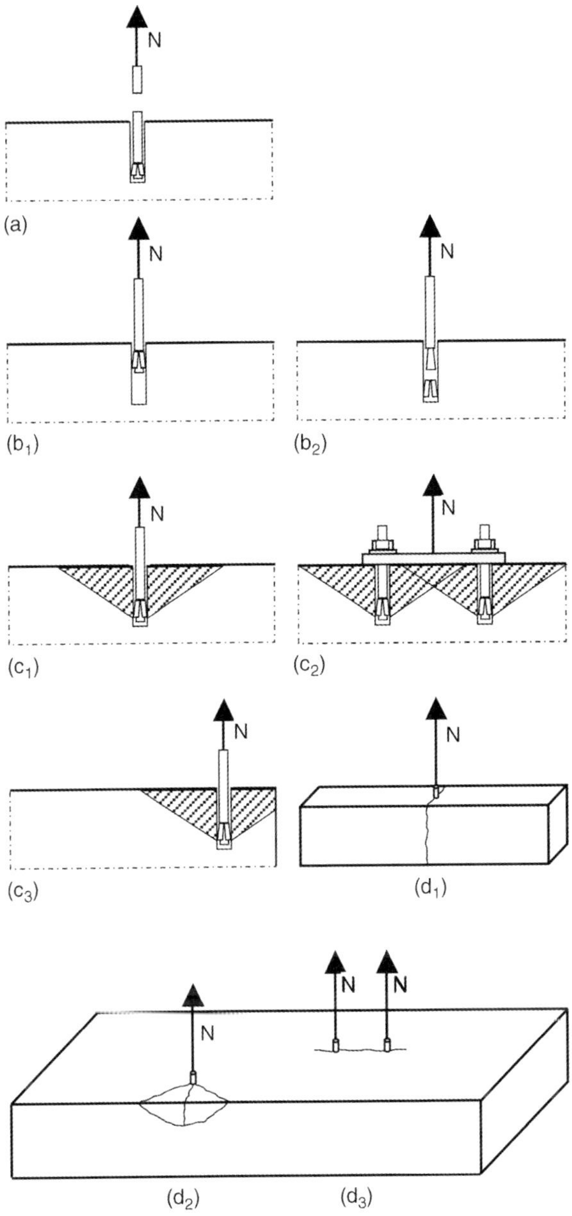

Fig. 5.1 Failure modes of post-installed fasteners under tension load
(a) steel failure
(b_1) pull-out failure
(b_2) pull-through failure
(c) conical concrete break-out failure
(d) splitting failure

Splitting failure is a failure mode in which the concrete component splits completely (Figure 5.1d_1) or the concrete fractures along the fastening and the edge of the concrete element (Figure 5.1d_2). In case of large distances to the edge splitting cracks might develop between closely spaced fasteners during installation (Figure 5.1d_3).

In general, it cannot be predicted which of the above mentioned failure modes will govern the load-bearing capacity of a fastening. Therefore all failure modes must be verified for design method *A*. In case of steel failure and pull-out or pull-through the resistance is neither influenced by neighbouring post-installed fasteners nor by component edges. For fastener groups the verification of these failure modes is performed for the most loaded single fastener. On the other hand spacing and edge distances are of major influence on the load-bearing capacity of fasteners in case of conical concrete break-out as well as splitting failure. Therefore, the verification shall be done for the whole fastener group if these failure modes occur.

5.2.2 Steel failure

The characteristic resistance of a post-installed fastener in case of steel failure is given in the relevant European Technical Approval. The strength calculations are based on f_{uk}. The steel resistance of a fastener is:

$$N_{Rk,s} = A_s \cdot f_{uk} \tag{5.1}$$

with:

A_s net cross section of the post-installed fastener, tensioned cross section for threaded parts

f_{uk} tensile steel strength

Bolt-type post-installed fasteners show a non-uniform cross section along the fastener length. In the upper area these post-installed fasteners have a thread with a smooth shaft following below. At the end of the bolt there is a conical segment with reduced cross section. All three areas can show different cross sections which for cold-formed fasteners might also indicate different tensile strengths. This is checked in the approval tests and the typical resistance given in the ETA is the smallest resistance of these three areas.

5.2.3 Pull-out/pull-through failure

The characteristic resistance in case of pull-out/pull-through failure is given in the relevant ETA. This value cannot be calculated but only determined in elaborate test series. In general, the value of the ETA is valid for the concrete strength class C20/25. For higher concrete strength classes (in general up to C50/60) the ETAs provide individual enlargement factors for each concrete strength class.

In the ETA it is noted, if no pull-out or pull- through failure appears. Then this verification does not have to be performed.

5.2.4 Conical concrete break-out failure

Most post-installed mechanical fasteners subjected to tension loads fail by cone-shaped concrete break-out. The break-out bodies of different post-installed fasteners are similar. The slope of the surfaces of the cones is not constant over the embedment depth and the circumference and varies from test to test. The average slope as measured from the horizontal lies between 30 ° and 40 ° and is on average about 35 °. As depth of the break-out cones the anchorage depth h_{ef} can be assumed as simplification. Then the diameter of a break-out cone is about three times the embedment depth of the post-installed fastener (Figure 5.2, on top).

The characteristic resistance of a group with post-installed fasteners failing by conical concrete break-out is influenced by numerous parameters. First there are the concrete strength and the condition of the concrete (cracked or uncracked) as well as the embedment depth of the post-installed fastener. Moreover, the spacing to neighbouring fasteners within a fastener group as well as the distances to the free edges of the concrete component affects the characteristic resistance. Finally a possible eccentricity of the load within a group as well as an unfavourable strong surface reinforcement

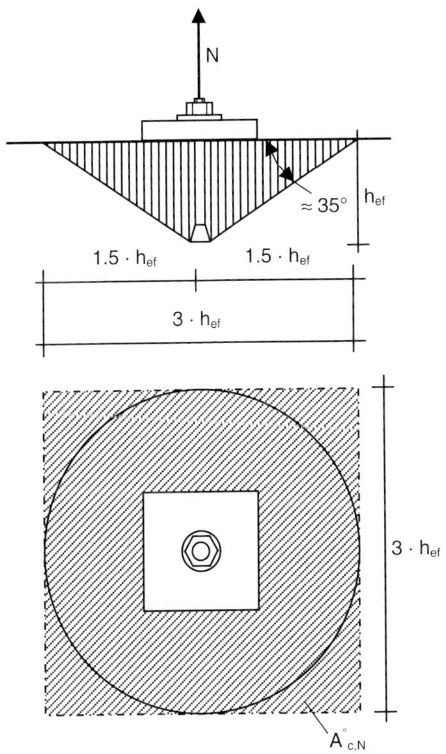

Fig. 5.2 Concrete break-out cone (schematically) (Fuchs, Eligehausen, and Breen (1995))

shall be taken into account. The characteristic resistance of a group with post-installed fasteners can be calculated after Equation 5.2:

$$N_{Rk,c} = N^0_{Rk,c} \cdot \frac{A_{c,N}}{A^0_{c,N}} \cdot \psi_{s,N} \cdot \psi_{re,N} \cdot \psi_{ec,N} \quad [\mathrm{N}] \tag{5.2}$$

The different factors of Equation 5.2 are explained in the following.

5.2.4.1 Characteristic resistance of a single fastener

Actually by use of realistic assumptions for the concrete characteristics it is possible to calculate numerically the load-bearing capacity of a single post-installed fastener with the failure mode concrete break-out with sufficient accuracy. Indeed, the efforts for these numerical investigations (e.g., calculations FEM) are very high. Hence, usually the resistance for concrete break-out failure is determined empirically on the basis of test results. The results of a very big number of test series and single tests form the basis of the approach for the characteristic resistance of a post-installed fastener with conical concrete break-out used in CEN/TS (Equations 5.2a$_1$ and 5.2a$_2$). Furthermore they consider the knowledge of the non-linear fracture mechanics (Rehm, Eligehausen, and Mallée (1992)). The equations are valid for single post-installed fasteners which are not influenced by neighbouring fasteners or component edges.

Cracked concrete:

$$N^0_{Rk,c} = k_{cr} \cdot \sqrt{f_{ck,cube}} \cdot h_{ef}^{1.5} \quad [\mathrm{N}] \tag{5.2a$_1$}$$

Uncracked concrete:

$$N^0_{Rk,c} = k_{ucr} \cdot \sqrt{f_{ck,cube}} \cdot h_{ef}^{1.5} \quad [\mathrm{N}] \tag{5.2a$_2$}$$

with:

k_{cr} product specific factor for cracked concrete
k_{ucr} product specific factor for uncracked concrete
$f_{ck,cube}$ characteristic concrete compressive strength [N/mm^2] for cubes with a side length of 150 mm considering the limitations given in the relevant ETA
h_{ef} embedment depth of the fastener [mm]

The product specific factors k_{cr} and k_{ucr} are given in the respective ETA. They differ not substantially for the individual types and sizes of post-installed fasteners. As a rule for mechanical post-installed fasteners they are $k_{cr} = 7.2$ und $k_{ucr} = 10{,}1$. The characteristic resistance in cracked concrete is lower than in uncracked concrete. This is caused mainly by the disturbance of the stress condition in the cracked concrete in the vicinity of the fastener. In uncracked concrete a tension load on a fastener generates a rotationally symmetric stress pattern around the fastener. Equilibrium is provided by the hoop stresses in the concrete (Figure 5.3a). If the fastening element is located in a wide crack, no tension forces can be transferred vertically to the crack. Hence, the crack causes a change of the stress distribution in the concrete (Figure 5.3b) and reduces the surface area available for the transfer of the tension forces. After this model one receives two independent concrete break-out bodies which are in contact in the area of the crack in the component.

Besides the described disturbance of the rotation-symmetrical stress condition in cracked concrete for undercut fasteners and displacement controlled fasteners the interlocking surface is decreased by the crack. In case of torque controlled expansion

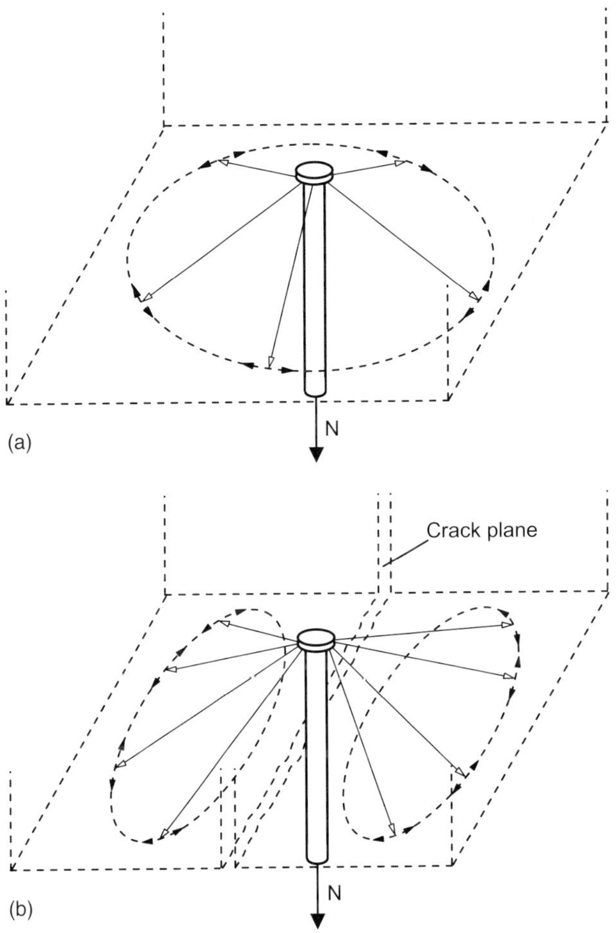

Fig. 5.3 Influence of a crack on the stress distribution in the area of a headed fastener under tension load
(a) uncracked concrete
(b) cracked concrete

fasteners the opening of the crack causes a reduction of the expansion forces. In case of post-installed expansion fasteners suitable for applications in cracked concrete, however, the reduction of the expansion forces is compensated by the fact that the expansion plug is pulled further into the expansion sleeves while the crack opens. This behaviour is called follow-up expansion.

5.2.4.2 Effect of spacing and edge distance

Spacing

The term $A_{c,N}/A^0_{c,N}$ in Equation 5.2 considers the effect of spacing and edge distances on the characteristic resistance for conical concrete break-out.

If fasteners within a group are installed with spacing that corresponds at least to the diameter of the break-out cone, then the break-out cones of neighboring fasteners (Figure 5.4a) do not overlap and the characteristic resistance of the group corresponds to n-times the resistance of a single fastener (n = number of the post-installed fasteners of the group). The corresponding characteristic spacing is $s_{cr,N}$ and amounts to $s_{cr,N} = 3 \cdot h_{ef}$. If the spacing is smaller than the characteristic value, the concrete break-out cones overlap (Figure 5.4b) and a common break-out body develops. Then the fractured surface available for the load introduction into the concrete is smaller than the sum of the fractured surfaces of the same number of single fasteners. Therefore the characteristic resistance decreases.

Theoretically it is possible to determine the effect of spacing within a fastener group on the characteristic resistance by comparison of the fractured surface of the group with the sum of the fractured surfaces of the same number of single fasteners. However, the calculation of overlapping cone surfaces is rather complex. Therefore, the approach described in CEN/TS assumes simplified and idealized concrete break-out bodies. It is based on a proposal by Fuchs (1991) and Fuchs, Eligehausen, and Breen (1995). The design method is called CC-method (Concrete Capacity Method) and has been adopted in many design provisions since then. It is based on κ – method which is described in Eligehausen, Mallée, and Rehm (1997).

The CC-method substitutes the break-out cone with a pyramid with the base length $s_{cr,N}$ and the height h_{ef}. For the idealized break-out body the base area $A^0_{c,N}$ on the concrete surface is quadratic (Figure 5.2). Is this area available on the surface of the component, a fastener reaches his maximum resistance for concrete break-out after Equation 5.2a$_1$ or Equation 5.2a$_2$. The influence of spacing on the characteristic concrete break-out resistance of a fastener group is taken into account by the factor $A_{c,N}/A^0_{c,N}$. Then the break-out resistance increases in proportion to the base area of the fastener group. Figure 5.5 shows examples for the determination of the base areas of the idealized break-out bodies for different fastener groups. In CEN/TS the base areas of one pyramid or the base areas of overlapping pyramids are called projected area.

In the following the CC-method is explained with the help of an example (Figure 5.6a). The embedment depth of the post-installed fasteners is $h_{ef} = 80$ mm. Then the characteristic spacing is $s_{cr,N} = 3 \cdot h_{ef} = 240$ mm. The projected area of the idealized break-out body of the group shown in Figure 5.6a amounts to $A_{c,N} = 171\,600$ mm^2 and

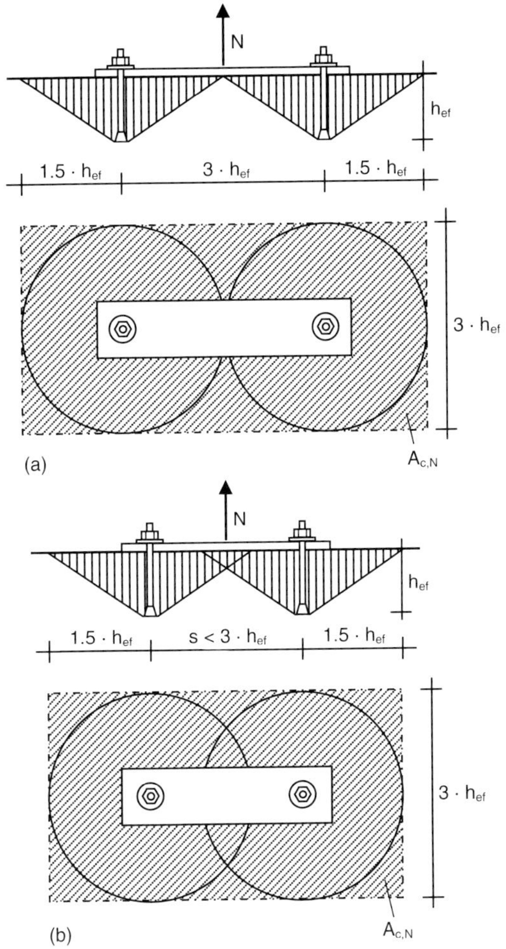

Fig. 5.4 Influence of fastener spacing on the break-out body of a double fastening
(a) spacing $s = 3 \cdot h_{ef}$
(b) spacing $s < 3 \cdot h_{ef}$

the projected area of a single post-installed fastener is $A^0_{c,N} = (s_{cr,N})^2 = 240^2 = 57\,600\ \text{mm}^2$. With this, the factor $A_{c,N}/A^0_{c,N} = 2.98$, that is the characteristic resistance of this fastener group is 2.98-times the value of the resistance of a single fastener. Therefore in this example the effect of the spacing in two directions compared with the sum of the resistance of 4 single fasteners reduces the resistance of the quadruple fastening by approximately 25%.

This approach is only valid if the spacing between the individual anchors of a group is less than $s_{cr,N} = 3 \cdot h_{ef}$. For greater spacing the break-out cones of adjacent fasteners do not overlap and the spacing is without effect on the resistance of the fastening.

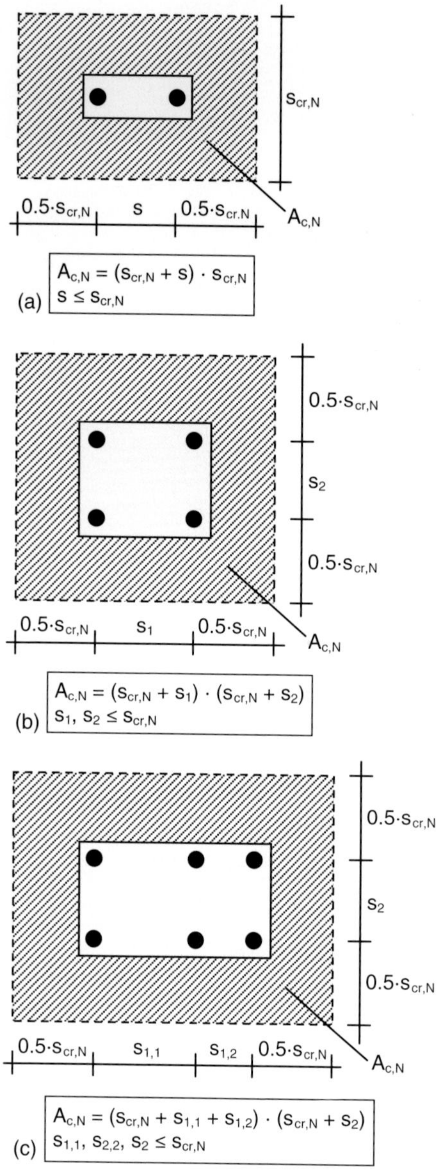

Fig. 5.5 Definition of the base area of the idealized concrete break-out body for different arrangements of fasteners

(a) double fastening

(b) quadruple fastening

(c) fixture with six fasteners and with different spacing $s_{1,1}$ and $s_{1,2}$

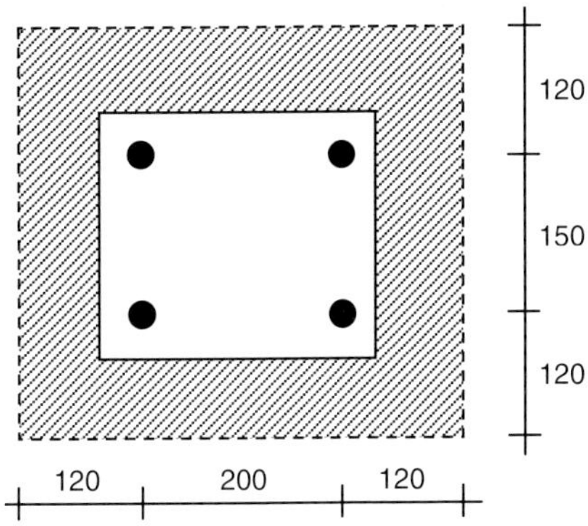

$$A_{c,N} = (0.5 \cdot s_{sr,N} + s_1 + 0.5 \cdot s_{sr,N}) \cdot (0.5 \cdot s_{sr,N} + s_2 + 0.5 \cdot s_{sr,N})$$
$$= (120 + 200 + 120) \cdot (120 + 150 + 120) = 171\,600 \text{ mm}^2$$

(a)

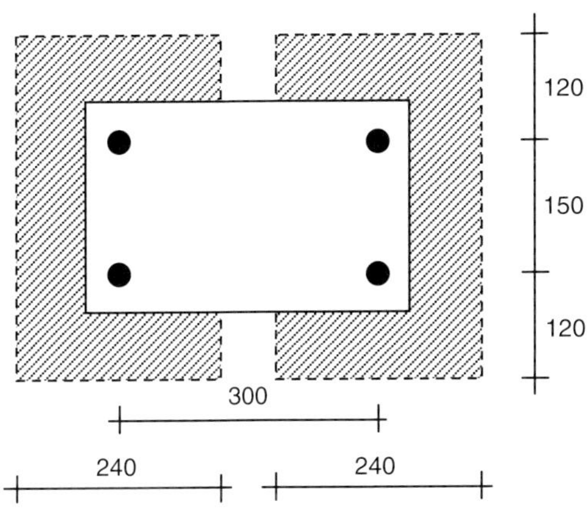

$$A_{c,N,1} = A_{c,N,2} = s_{sr,N} \cdot (0.5 \cdot s_{sr,N} + s_2 + 0.5 \cdot s_{sr,N})$$
$$= 240 \cdot (120 + 150 + 120)$$
$$= 93\,600 \text{ mm}^2$$

(b)

Fig. 5.6 Determination of the base area of the idealised concrete break-out body at the example of a quadruple fastening (embedment depth $h_{ef} = 80$ mm)
(a) spacing s_1 and s_2 smaller than $s_{cr,N} = 240$ mm
(b) spacing $s_1 > s_{cr,N} = 240$ mm

An example is given in Figure 5.6b. The horizontal spacing is $s_1 = 300$ mm. This value is greater than the characteristic spacing $s_{cr,N} = 240$ mm. This means that the base plate is fastened by two double fastenings which act independently. The base areas of each of the two double fastenings is $A_{c,N} = 93\,600\text{ mm}^2$ and $A^0_{c,N} = 240^2 = 57\,600\text{ mm}^2$. The ratio of the base areas is $A_{c,N}/A^0_{c,N} = 1.63$, that is, the characteristic resistance of each double fastenings is 1.63-times the value of a single fastening.

A simple and easy to understand geometrical model which can theoretically be used to design a myriad of applications with post-installed fasteners even in cases, which are beyond the area of application of CEN/TS (Figure 2.4) forms the basis of the CC method. Its validity was verified for groups with up to 36 fasteners (Eligehausen *et al.* 1992)). It is valid also for groups with variable spacing. Figure 5.7 shows examples. In Figure 5.7a all spacing of the group are smaller than the characteristic value $s_{cr,N}$; in Figure 5.7b, however, the spacing $s_{1,2}$ is bigger, that is, the base plate is fixed by two separate fastener groups.

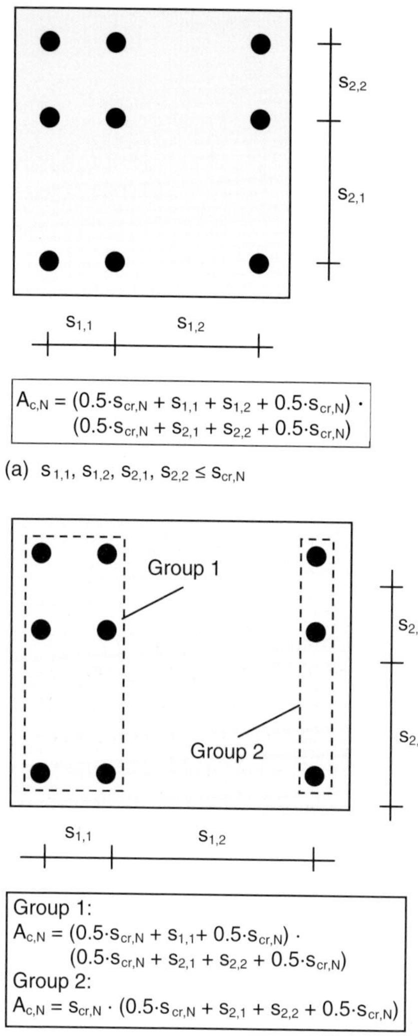

(a) $s_{1,1}, s_{1,2}, s_{2,1}, s_{2,2} \leq s_{cr,N}$

(b) $s_{1,1}, s_{2,1}, s_{2,2} \leq s_{cr,N}$; $s_{1,2} > s_{cr,N}$

Fig. 5.7 Examples of fastener groups with different spacing

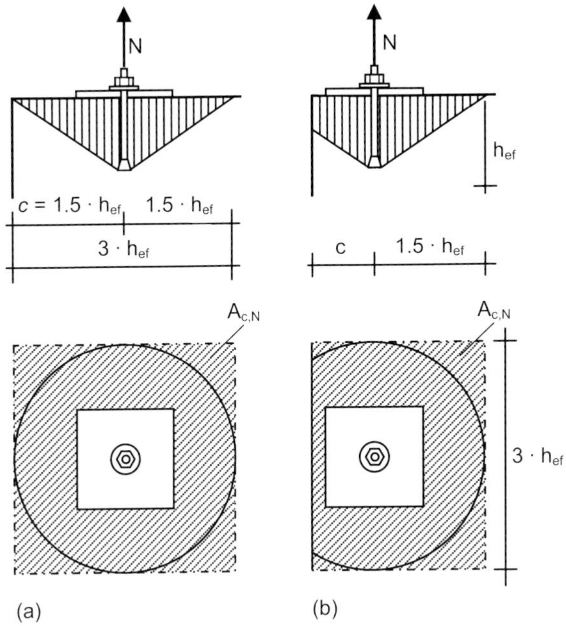

Fig. 5.8 Influence of a component edge on the shape of the concrete break-out cone (Fuchs, Eligehausen, and Breen (1995))
(a) fastening with an edge distance $c = 1.5 \cdot h_{ef}$
(b) fastening with an edge distance $c < 1.5 \cdot h_{ef}$

Edge distances

The geometrical influence of component edges on the characteristic resistance with conical concrete break-out (Figure 5.8) can be also described with the help of the CC-method. If the edge distance of a fastener corresponds to the radius of the failure cone, the cone is tangent to the edge (Figure 5.8a) and the load-carrying capacity corresponds to the value of a one single fastener. If the edge distance is reduced (see Figure 5.8b), then failure cone and edge overlap and the characteristic resistance decreases. The influence can be described – as in case of spacing – by comparison of the base area of the idealized (truncated) break-out body with the base area of a single fastener which is not influenced by spacing or edge distances. Figure 5.9 explains the approach at the example of a single post-installed fastener at the component edge (Figure 5.9a) and a quadruple fastening positioned in the corner of a component (Figure 5.9b).

Figure 5.10 shows the approach for a group with two post-installed fasteners in the corner of a component. In the example post-installed fasteners with an anchorage depth $h_{ef} = 80$ mm are assumed. In Figure 5.10a the spacing is $s < s_{cr,N}$ and in Figure 5.10b the spacing is larger. In the first case the break-out cones of both fasteners overlap and in the second they are separate.

The geometrical influence of the overlapping of the failure cones with component edges is described by the comparison of the base areas. In addition, it is to be noted, that

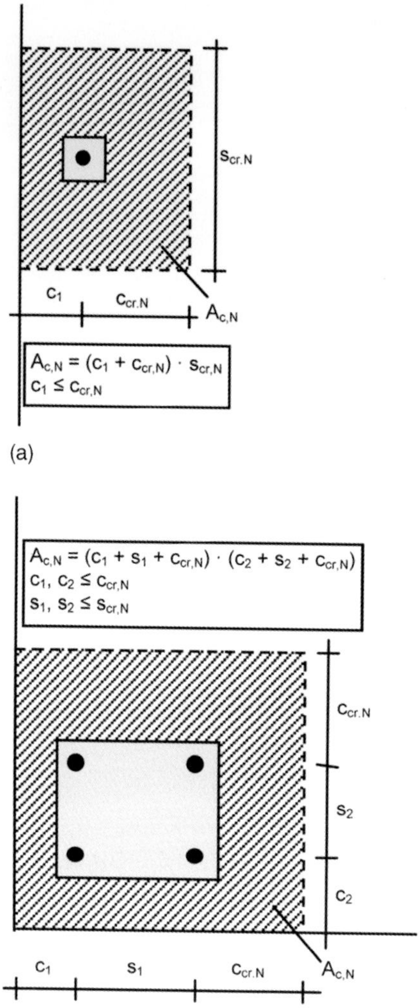

Fig. 5.9 Definition of the base area of the idealized concrete break-out body of different fastenings close to edge
(a) single fastener at the component edge
(b) quadruple fastening in the corner of a component

the condition of stress in the area of a post-installed fastener is disturbed – similar to cracks in the base material – also by edge distances. Figure 5.11 shows this influence. In Figure 5.11a a tension loaded headed fastener is shown without edge influence. The stresses are distributed rotation-symmetrically. A component edge can be compared to a crack which is so wide that no tensile stresses can be transferred. In Figure 5.11b the

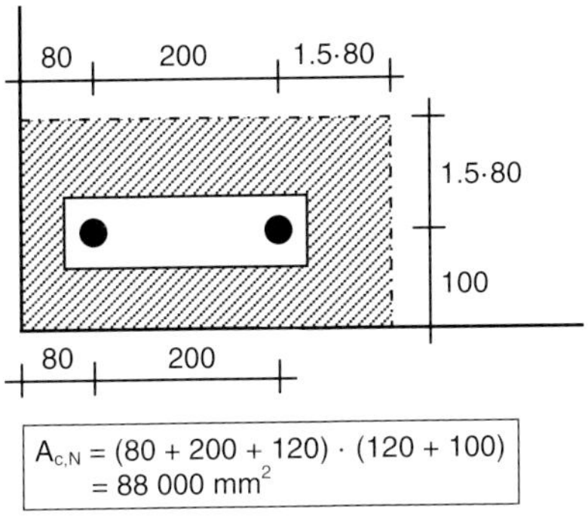

(a)

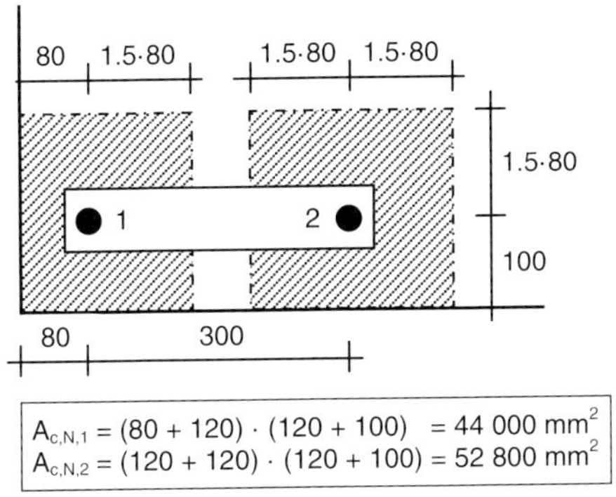

(b)

Fig. 5.10 Determination of the base area of the idealized concrete break-out body at the example of a double fastening located at the edge of a component (embedment depth $h_{ef} = 80$ mm)
(a) spacing $s \leq s_{cr,N}$
(b) spacing $s > s_{cr,N}$

resulting change of the stress distribution is shown. This influence is considered in Equation 5.2 by the factor $\psi_{s,N}$.

$$\psi_{s,N} = 0.7 + 0.3 \cdot \frac{c}{c_{cr,N}} \leq 1.0 \qquad (5.2b)$$

If more than one edge is involved (e.g. a fastening in a corner or in a narrow member), the smallest edge distance is used in Equation 5.2b.

5.2.4.3 Effect of heavy surface reinforcement (shell spalling)

Up to now it was assumed that the post-installed fasteners are positioned in unreinforced concrete. Nevertheless, in practice structural concrete components are mostly reinforced. The presence of orthogonal reinforcement near the surface of slabs and walls does not

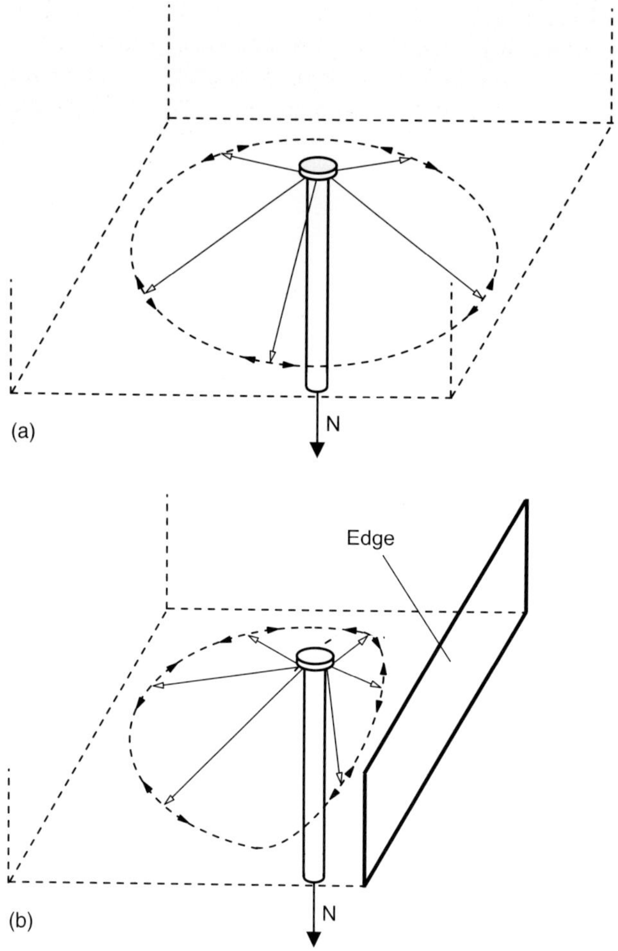

Fig. 5.11 Influence of a component edge on the stress distribution in the area of a headed fastener under tension load
(a) fastening without edge influence
(b) fastening in the component edge

increase the tension capacity of fasteners with conical concrete break-out because it is oriented vertically to the direction of the acting force. However, the reinforcement can cause a more ductile post-peak-failure behaviour if the break-out cone is supported by the wire mesh. This occurs only if the reinforcement is closely spaced and is further engaged by stirrups with small spacing (Rehm and Pusill-Wachtsmuth (1979)).

An unfavourable condition with respect to the concrete break-out resistance is created when post-installed fasteners are located within the concrete cover or close to dense reinforcement. The bond stresses associated with the reinforcing bars are superimposed on the tensile stresses generated by the fasteners. Additionally, the closely spaced reinforcement disrupts the transfer of the force from the fastener into the member ('perforated' concrete cover). Aggravating this condition is the fact that the concrete strength in the concrete cover is often lower and the concrete is of poorer quality than that in the middle of the member cross-section. These effects are considered by the shell spalling factor $\psi_{re,N}$ (Eligehausen *et al.* 1989)):

$$\psi_{re,N} = 0,5 + \frac{h_{ef}}{200} \leq 1,0 \tag{5.2c}$$

with:

h_{ef} embedment depth of the post-installed fastener [mm]

The shell spalling factor may be taken $\psi_{re,N} = 1.0$ in the following cases:

- reinforcement (any diameter) is provided at a spacing $\geq$150 mm, or
- reinforcement with a diameter of 10 mm or less is provided at a spacing $\geq$100 mm.

5.2.4.4 Effect of the eccentricity of the load

Bending moments acting on a fastener group stress the individual fasteners of the group with different tension loads. This was not considered in the foregoing discussions. The effect of load eccentricities is covered based on a proposal by (Riemann (1985)) by means of the factor $\psi_{ec,N}$ which takes account of a group effect when different tension loads are acting on the individual fasteners of a group. This approach was developed based on the verification of punching shear of flat slabs by Moe (1961). It is formed on the following analogy:

If a flat slab is supported on a centrically loaded round column, in the failure state this column punches a cone out of the ceiling. Analogously in the state of failure a fastening under concentric loading pulls out a cone from the concrete. If the round column is loaded eccentrically, the failure cone is punched eccentrically from the ceiling due to the moment acting on the head of the column. Analogously an eccentrically loaded fastening in the rupture state tears out the cone eccentrically from the concrete. Based on this analogy the eccentricity factor $\psi_{ec,N}$ can be calculated as follows:

$$\psi_{ec,N} = \frac{1}{1 + 2 \cdot e_N / s_{cr,N}} \leq 1,0 \tag{5.2d}$$

with:

e_N eccentricity of the resulting tensile load acting on the tensioned fasteners to the geometric centroid of the tension-loaded fasteners

Where there is an eccentricity in two directions then Equation 5.2d is used to calculate $\psi_{ec,N}$ for each axis separately and the product of both factors is used in Equation 5.2.

Examples for the calculation of the eccentricity of the resulting tension load of fasteners are shown in Figure 5.12. In the example of Figure 5.12a only the right

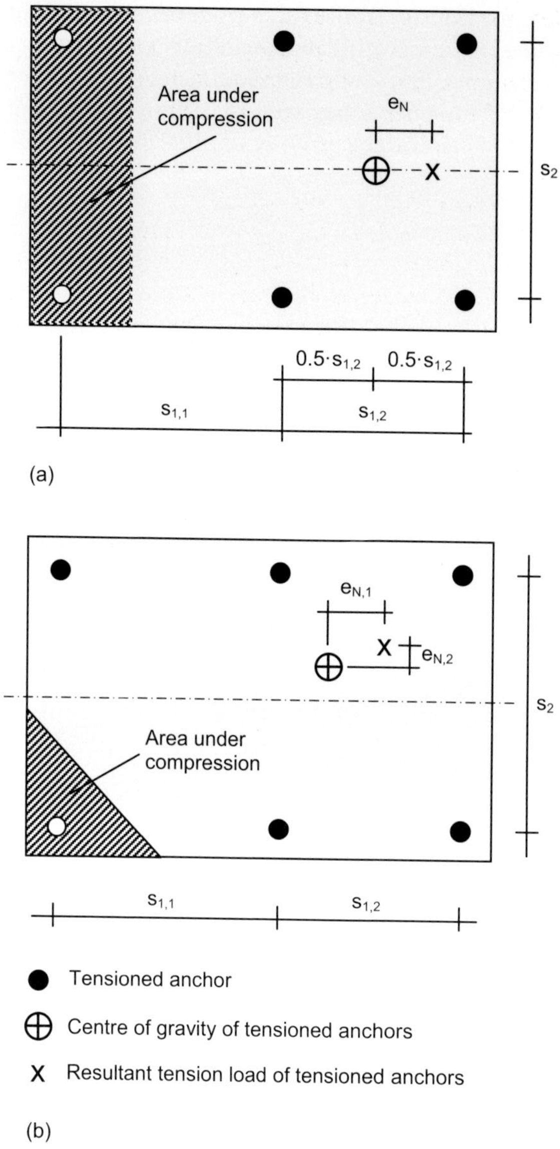

Fig. 5.12 Examples of the eccentricity of the resultant tension load (Eligehausen, Mallée, and Rehm (1997))
(a) uni-axail bending
(b) oblique bending

four fasteners are loaded. Then the eccentricity e_N of the resultant tensile force of these four fasteners is derived with respect to the centre of gravity of the tension-loaded fasteners only. Figure 5.12b shows an eccentricity about two axes. The left lower fastener lies in the compression zone below the base plate and is not tensioned. The eccentricities $e_{N,1}$ and $e_{N,2}$ are determined for the remaining five post-installed fasteners. In this application, for reasons of simplicity, the group of tensioned anchors may be resolved into a group rectangular in shape, that means the centre of gravity of the tensioned anchors may be assumed in the centre of gravity of the resolved group. This approach is conservative, because it yields larger eccentricities.

5.2.4.5 Special cases: three or four edges with $c_i < c_{cr,n}$

In presence of three or four edges with $c_i < c_{cr,N} < 1.5 \cdot h_{ef}$ (Figure 5.13) Equation 5.2 leads to conservative results. This is illustrated with the example in Figure 5.14. Figure 5.14 shows a single tension-loaded fastener embedded in a concrete prism with equal distance $c_i = 100$ mm to the edge. The component of concrete strength class C20/25 is uncracked. In the first case the embedment depth of the fastener is $h_{ef} = 100$ mm and in the second case $h_{ef} = 200$ mm. Hence, the characteristic resistance as predicted by Equation 5.2 becomes:

Case 1:

$$N_{Rk,c} = 10.1 \cdot \sqrt{25} \cdot 100^{1.5} \cdot \frac{40\,000}{90\,000} \cdot \left(0.7 + 0.3 \cdot \frac{100}{150}\right) \cdot 1.0 \cdot 1.0 = 20\,200 \quad [\text{N}]$$

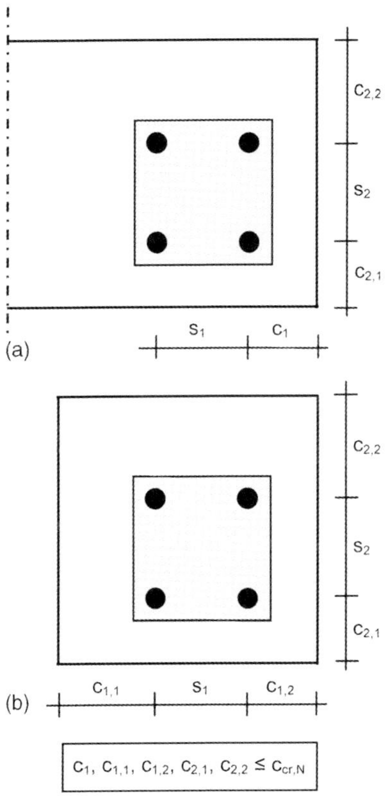

Fig. 5.13 Examples of the special case in which the modified embedment depth h'_{ef} may be applied
(a) quadruple fastening with three component edges
(b) quadruple fastening with four component edges

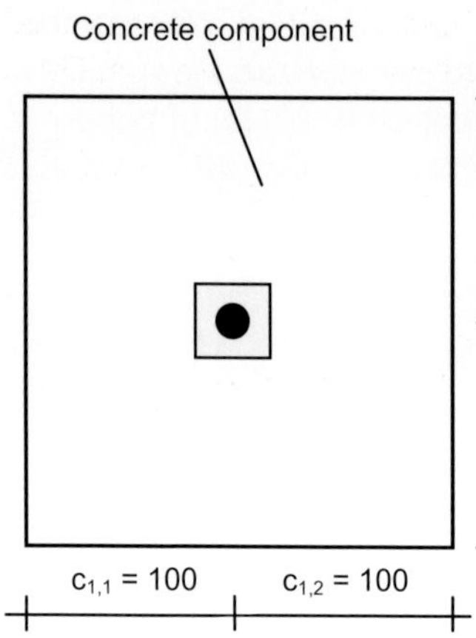

Fig. 5.14 Single fastener influenced by four component edges

Case 2:

$$N_{Rk,c} = 10.1 \cdot \sqrt{25} \cdot 200^{1.5} \cdot \frac{40\,000}{360\,000} \cdot \left(0.7 + 0.3 \cdot \frac{100}{300}\right) \cdot 1.0 \cdot 1.0 = 12\,700 \quad [\mathrm{N}]$$

The characteristic resistance is in the second case smaller than in the first one in spite of double embedment depth. This is not logical. In reality, regardless of the embedment depth of the fastener the concrete prism fractures always in the effective anchorage zone. More accurate results are obtained if in all applications with three or four edges the embedment depth h_{ef} is replaced by the larger result for h'_{ef} of Equation 5.2e:

$$h'_{ef} = \frac{c_{max}}{c_{cr,N}} \cdot h_{ef} \tag{5.2e$_1$}$$

$$h'_{ef} = \frac{s_{max}}{s_{cr,N}} \cdot h_{ef} \tag{5.2e$_2$}$$

Where: c_{max} is the largest edge distance ($\leq c_{cr,N}$) and s_{max} the largest spacing within a group ($\leq s_{cr,N}$). Then, the characteristic spacing and edge distance are:

$$s'_{cr,N} = \frac{h'_{ef}}{h_{ef}} \cdot s_{cr,N} \tag{5.2f$_1$}$$

$$c'_{cr,N} = \frac{h'_{ef}}{h_{ef}} \cdot c_{cr,N} \tag{5.2f$_2$}$$

The calculation of the characteristic resistance with these modified characteristic spacing results in the same characteristic resistances for both cases of Figure 5.14:

$$h'_{ef} = \frac{100}{150} \cdot 100 = 66.67 \quad [\text{mm}] \quad \text{or} \quad h'_{ef} = \frac{100}{300} \cdot 200 = 66.67 \quad [\text{mm}]$$

$$s'_{cr,N} = \frac{66.67}{100} \cdot 300 = 200 \quad [\text{mm}] \quad \text{or} \quad s'_{cr,N} = \frac{66.67}{200} \cdot 600 = 200 \quad [\text{mm}]$$

$$c'_{cr,N} = \frac{66.67}{100} \cdot 150 = 100 \quad [\text{mm}] \quad \text{or} \quad c'_{cr,N} = \frac{66.67}{200} \cdot 300 = 100 \quad [\text{mm}]$$

$$N_{Rk,c} = 10.1 \cdot \sqrt{25} \cdot 66,67^{1.5} \cdot \frac{40\,000}{40\,000} \cdot \left(0.7 + 0.3 \cdot \frac{100}{100}\right) \cdot 1.0 \cdot 1.0 = 27\,490 \quad [\text{N}]$$

The calculated characteristic resistance represents the maximum value of the geometry under consideration. If the embedment depth h_{ef} is reduced the characteristic resistance decreases and if it is increased the characteristic resistance remains constant. The modification of the embedment depth used in the calculation resolves this weak point of the design method.

The values h'_{ef}, $S'_{cr,w}$ and $C'_{cr,w}$ are used for the determination of the areas $A_{c,N}$ and $A^0_{c,N}$ and inserted in Equations 5.2a$_1$, 5.2a$_2$, 5.2b and 5.2d.

5.2.5 Splitting

5.2.5.1 Splitting failure during installation of post-installed fasteners

Splitting failure during installation of post-installed fasteners does not occur, if the minimum values for edge distance (c_{min}), spacing (s_{min}) and member thickness (h_{min}) required by the ETA are observed.

5.2.5.2 Splitting failure of loaded post-installed fasteners

Splitting failure of post-installed fasteners subjected to tension loads occurs only, if the fasteners are positioned close to the edge. Fastenings located remote from the edge ('in the field') are not able to entirely split the entire component. On the other hand, a suitable reinforcement can compensate for the effects of splitting cracks. Therefore, CEN/TS requires no verification of splitting if at least one of the following conditions is fulfilled:

- the edge distance in all directions is $c \geq c_{cr,sp}$ for single fasteners and $c \geq 1.2 \cdot c_{cr,sp}$ for fastener groups. The value $c_{cr,sp}$ is given in the relevant ETA.
- the characteristic resistance for concrete cone failure and pull-out failure is calculated for cracked concrete and reinforcement located in the anchorage zone of the fastener resists the splitting forces and limits the crack width to $w_k \leq 0.3$.

If none of the above conditions is fulfilled, then splitting failure has to be verified. The literature indicates several proposals to the subject 'splitting'. Their area of application is limited, however, since the validity is limited either to certain types of fasteners or applications or they are less suitable for design purposes. Therefore, in present design provisions and in CEN/TS a simplistic design procedure which is based essentially on the approach of conical concrete break-out is suggested. This analogy makes sense, because the resistance of a fastener causing splitting failure is influenced by the same parameters as in case of concrete break-out. Thus, for the resistance of a single post-installed fastener the concrete strength is of influence and the size of the cracked surface is influenced by the embedment depth as well as by the spacing of fasteners in a group and edge distances. Also the load eccentricity affects the resistance of splitting. On the other hand, the concrete break-out resistance is independent of the component thickness, in case of splitting, however, this influence exists. The thicker the component is, the greater the force necessary to split the component. Therefore, the thickness of the concrete member must be considered in the verification of splitting.

With the above assumptions the characteristic resistance of a group with post-installed fasteners can be determined:

$$N_{Rk,sp} = N_{Rk}^{0} \cdot \frac{A_{c,N}}{A_{c,N}^{0}} \cdot \psi_{s,N} \cdot \psi_{re,N} \cdot \psi_{ec,N} \cdot \psi_{h.sp} \quad [\mathrm{N}] \tag{5.3}$$

N_{Rk}^{0} is the minimum of the resistances $N_{Rk,p}$ and $N_{Rk,c}^{0}$. The characteristic resistance $N_{Rk,p}$ for pull-out is given in the relevant ETA and the resistance of a single post-installed fastener failing by conical concrete break-out is calculated by Equation 5.2a$_1$ for cracked concrete and Equation 5.2a$_2$ for uncracked concrete.

The values $\psi_{s,N}$, $\psi_{re,N}$ und $\psi_{ec,N}$ as well as the areas $A_{c,N}$ und $A_{c,N}^{0}$ are calculated with Equations 5.2b, 5.2c and 5.2d as well as can be determined analogous to Figure 5.5 where the values $s_{cr,N}$ and $c_{cr,N}$ are replaced by $s_{cr,sp}$ and $c_{cr,sp}$. The characteristic spacing $s_{cr,sp}$ and edge distance $c_{cr,sp}$ for splitting are given in the relevant ETA.

The factor $\psi_{h,sp}$ considers the above mentioned effect of the member thickness. It is given by Equation 5.3a:

$$\psi_{h.sp} = \left(\frac{h}{h_{min}}\right)^{2/3} \leq \left(\frac{2 \cdot h_{ef}}{h_{min}}\right)^{2/3} \tag{5.3a}$$

The value $h_{\min}$ in Equation 5.3a corresponds to the minimum member thickness required by the ETA.

If the edge distance is smaller than the characteristic value $c_{cr,sp}$ a longitudinal reinforcement should be provided along the edge of the concrete member at the height of the anchorage zone of the fastener.

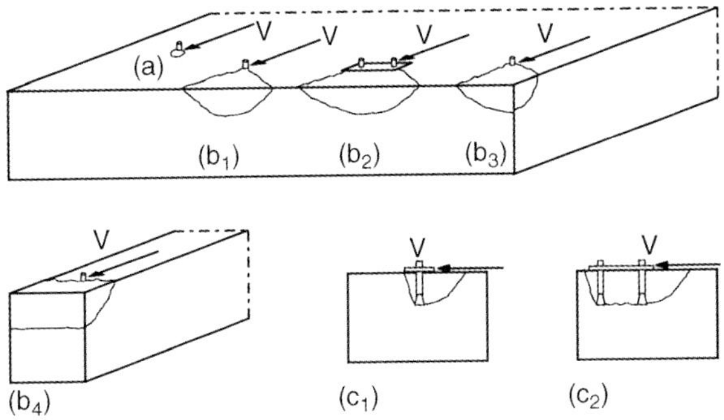

Fig. 5.15 Failure modes of fasteners under shear load
(a) steel failure
(b) concrete edge failure
(c) concrete pry-out failure

5.3 Shear load

5.3.1 Required verifications

Post-installed fasteners loaded in shear with and without lever arm exhibit steel failure, concrete pry-out and concrete edge failure (Figure 5.15). Steel failure due to shear loading without lever arm is defined as rupture of the steel stud, bolt or sleeve (Figure 5.15a). It represents the upper limit on the achievable resistance of a fastener in shear. In general, shell shaped spalling of the surface concrete precedes steel failure. However, this spalling does not influence the peak load but only the displacement of the post-installed fastener up to steel failure.

Fasteners having limited embedment and large diameters can exhibit sufficient rotation to produce a pry-out failure. Then the fracture surface develops 'behind' the point of load application (Figure 5.15c_1 and c_2). After closer analysis this failure can be considered a tension failure of the fasteners (Figure 5.16). With increasing shear load the surface concrete is damaged caused by the high pressures in the borehole mouth (shell-shaped spalling) shifting the resultant V_b of the concrete pressure along the fastener to a location deeper in the concrete. At the same time the baseplate rotates and loses contact with the concrete on the loaded side. These two mechanisms act to further increase the eccentricity between the applied shear load V and the concrete stress resultant V_b in the concrete. The moment originating this eccentricity generates a compressive force C between baseplate and concrete and a tensile force N in the fastener. If the tensile force in the fastener exceeds the tensile capacity associated with the maximum fracture surface, that can be activated by the fastener, a fracture surface originating at the lower end of the fastener forms behind the fastener.

Fasteners located close to the edge fail by concrete edge break-out (Figure 5.15b_1). A group of fasteners loaded in shear and proximate to an edge may develop a common fracture surface (Figure 5.15b$_2$), and the development of the fracture surface is interrupted by the presence of a corner (Figure 5.15b$_3$), or by proximate edges parallel with the load direction that is by a narrow component (Figure 5.15b$_4$).

As with fastenings under tension load it also cannot be predicted under shear load, which of the mentioned failure modes is governing the load-carrying capacity of a fastening. Therefore, all failure modes must be verified in design method *A*. The resistance for steel failure is not influenced by neighbouring post-installed fasteners or by component edges. In this case verification is performed for the highest loaded fastener. On the other hand spacing and edge distances significantly influence the load-carrying capacity of fasteners in case of concrete pry-out failure and concrete edge break-out. Therefore, with these failure modes the entire post-installed fastener group is to be verified.

5.3.2 Steel failure without lever arm

For post-installed fasteners the characteristic resistance of a shear-loaded fastener without lever arm in case of steel failure is given in the relevant ETA. The strength calculations are based on f_{uk}:

$$V_{Rk,s} = 0.5 \cdot A_s \cdot f_{uk} \tag{5.4}$$

with:

A_s net cross section of the post-installed fastener, tensioned cross section for threaded parts

f_{uk} tensile steel strength

Post-installed fasteners with relatively low ductility exhibit small displacements at failure. Therefore the load-carrying capacity of fastener groups with a diameter of the hole in the base plate d_f after Table 4.1 is reduced. Hence, the characteristic resistance $V_{Rk,s}$ given in the relevant ETA should be multiplied with the factor 0.8.

Wedge type or bolt type post-installed fasteners have a non-uniform cross section along the length of the fastener. Therefore, during the prequalification procedure of the fastener the cross section governing the failure (thread, shaft or cone) is determined in tests. The governing characteristic resistance can be taken from the relevant ETA.

A comparison of the characteristic steel resistance in tension and shear indicates (Equations 5.1 and 5.4), that the shear resistance is smaller than the tension resistance. This is due to the fact that the big deformations under shear loading cause superimposition of shear, bending and tensile stresses at failure state.

5.3.3 Steel failure with lever arm

In stand-off installations (Figure 2.5d) or applications with a grout layer with a thickness $>0{,}5 \cdot d$ between fixture and concrete surface or if the compressive strength of the grout $<30\ \mathrm{N/mm^2}$ a shear load with lever arm has to be observed (see

Section 4.3.1). Then, the characteristic steel resistance may be obtained from Equation 5.5:

$$V_{Rk,s} = \frac{\alpha_M \cdot M_{Rk,s}}{l} \tag{5.5}$$

with:

α_M see Section 4.3.1
l see Section 4.3.1
$M_{Rk,s}$ characteristic bending moment

$$= M^0_{Rk,s} \cdot (1 - N_{Ed}/N_{Rd,s}) \tag{5.5a}$$

$M^0_{Rk,s}$ characteristic bending moment of a single fastener, see ETA
N_{Ed} tension load acting on the fastener

$$N_{Rd,s} = N_{Rk,s}/\gamma_{Ms} \tag{5.5b}$$

Equation 5.5a considers that the fastener under consideration might be simultaneously subjected to a tension force, already taking a share of the useable steel strength. With increasing share of the tensile stress the share of the bending moment decreases in relation. If the tension force uses entirely the available steel strength, then $N_{Ed}/N_{Rd,s} = 1$ and no additional bending moment can be resisted.

The design provisions for shear loads with lever arm of CEN/TS are conservative for fastenings with grout layers. A more realistic design model is described in Fichtner (2011).

5.3.4 Pry-out failure

In case of concrete pry-out the fasteners fail due to the tension force generated by the displacements of the shear-loaded post-installed fasteners (Figure 5.16). At ultimate it

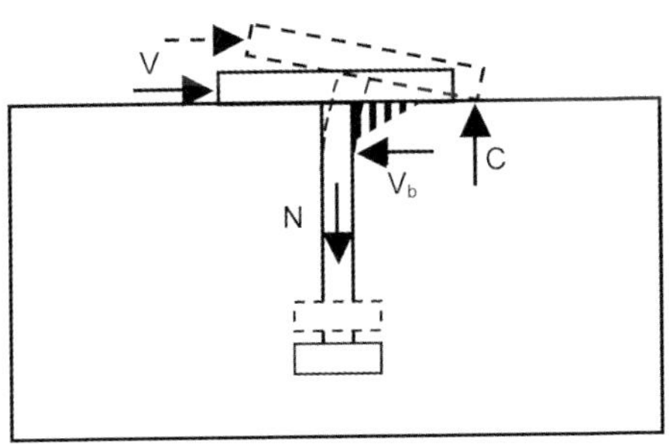

Fig. 5.16 Failure mechanism of a headed fastener in case of concrete pry-out (Zhao (1993))

is approximately about 30% to 50% of the applied shear load (Fuchs (1990)). The break-out body is smaller than a conical concrete break-out. The peak load, however, is affected by the same parameters. Therefore the characteristic resistance in case of pry-out failure can be calculated from the value for conical concrete break-out of fastenings in tension (Zhao (1993)):

$$V_{Rk,cp} = k_3 \cdot N_{Rk,c} \tag{5.6}$$

with:

k_3 see ETA
$N_{Rk,c}$ according to Equation 5.2, calculated for the fasteners resisting shear forces

According to current experience $k_3 = 1$ for post-installed fasteners with $h_{ef} < 60\,\text{mm}$ and $k_3 = 2$ for post-installed fasteners with $h_{ef} \geq 60\,\text{mm}$.

The following is to be observed:

In a fastener group subjected to a tension force and a bending moment, individual fasteners might be located in the compression below the base plate and not be loaded in tension. In case of pry-out failure, however, it is assumed that all fasteners within a group resist shear forces if no slotted holes are present and the hole diameter d_f in the base plate does not exceed the values of Table 4.1 (see Section 4.3.1). This is shown in Figure 5.17a with a quadruple fastening subjected to a normal force and a bending moment. The two fasteners on the left are situated in the compression zone below the base plate and, hence, are not loaded in tension. On the other hand all four post-installed fasteners resist the shear force. If only the resistance $N_{Rk,c}$ of both tension loaded post-installed fasteners is assumed in Equation 5.6 the result is conservative because in reality all four post-installed fasteners take up shear loads. Are single post-installed fasteners of the group located in slotted holes, not all tension loaded fasteners are loaded by a shear force. This is shown in Figure 5.17b. The quadruple fastening is loaded by a concentric working tension force (all fasteners take up tension loads), while because of the slotted holes only both post-installed fasteners on the right are under shear load. If one takes into account, however, the resistance of the four tension loaded post-installed fasteners in this case, the result is unambiguously liberal. Therefore, not the characteristic resistance $N_{Rk,c}$ for conical concrete break-out calculated for the verification in tension should be assumed in both cases in Equation 5.6, but be calculated new for the shear-loaded post-installed fasteners.

Furthermore it is to be noted that in Equation 5.2d not the eccentricity e_N of the tension loaded post-installed fasteners is to be inserted, but the eccentricity of the resultant of the shear loads to the centre of gravity of the shear-loaded post-installed fasteners.

If a fastener group is loaded by a torsion moment, the direction of the shear loads acting on the post-installed fasteners changes. An example is given in Figure 5.18. In this case the sum of the shear forces acting on the post-installed fasteners is zero and

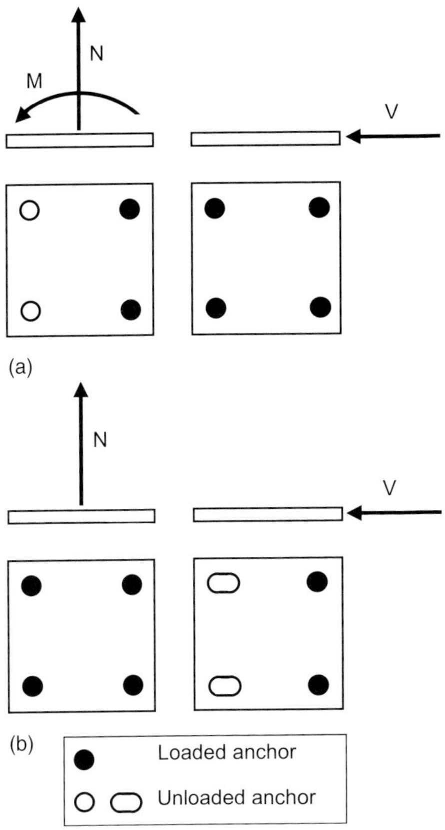

Fig. 5.17 Applications in which different fasteners must resist tension and shear forces
(a) loading of fastening by normal force, bending moment and shear load, fastener in round holes
(b) loading of fastening by normal force and shear load, fasteners in some cases in slotted holes

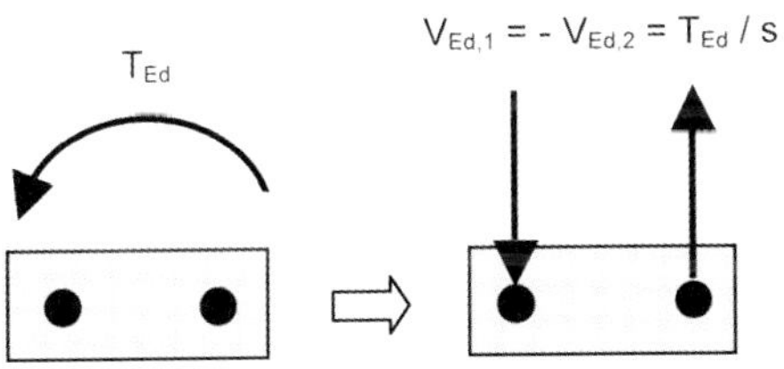

Fig. 5.18 Double fastening, loaded by a torsion moment

Equation 3.1 cannot be used. Therefore, CEN/TS requires a verification of concrete pry-out of the most unfavourable fastener in all cases in which vertical ($V_{Ed,v}$) or horizontal components ($V_{Ed,h}$) of shear loads on the post-installed fasteners of the group change their direction. Figure 5.19 shows examples of the calculation of the base area of the idealized break-out body.

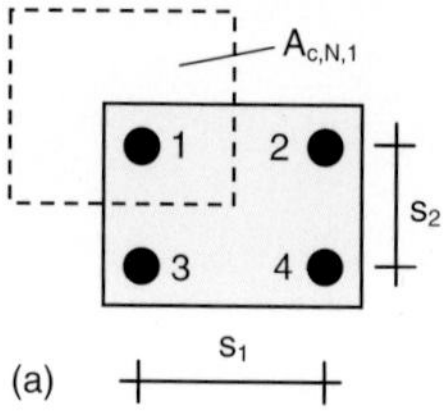

(a)

$$A_{c,N,1} = (0.5 \cdot s_{cr,N} + 0.5 \cdot s_1) \cdot (0.5 \cdot s_{cr,N} + 0.5 \cdot s_2)$$
$$A_{c,N,2} = A_{c,N,1}$$
$$A_{c,N,3} = A_{c,N,1}$$
$$A_{c,N,4} = A_{c,N,1}$$
$$s_1,\ s_2 \leq s_{cr,N}$$

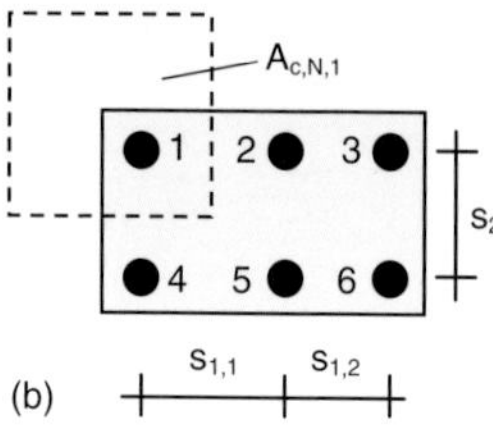

(b)

$$A_{c,N,1} = (0.5 \cdot s_{cr,N} + 0.5 \cdot s_{1,1}) \cdot (0.5 \cdot s_{cr,N} + 0.5 \cdot s_2)$$
$$A_{c,N,2} = (0.5 \cdot s_{1,1} + 0.5 \cdot s_{1,2}) \cdot (0.5 \cdot s_{cr,N} + 0.5 \cdot s_2)$$
$$A_{c,N,3} = (0.5 \cdot s_{1,2} + 0.5 \cdot s_{cr,N}) \cdot (0.5 \cdot s_{cr,N} + 0.5 \cdot s_2)$$
$$A_{c,N,4} = A_{c,N,1}$$
$$A_{c,N,5} = A_{c,N,2}$$
$$A_{c,N,6} = A_{c,N,3}$$
$$s_{1,1},\ s_{1,2},\ s_2 \leq s_{cr,N}$$

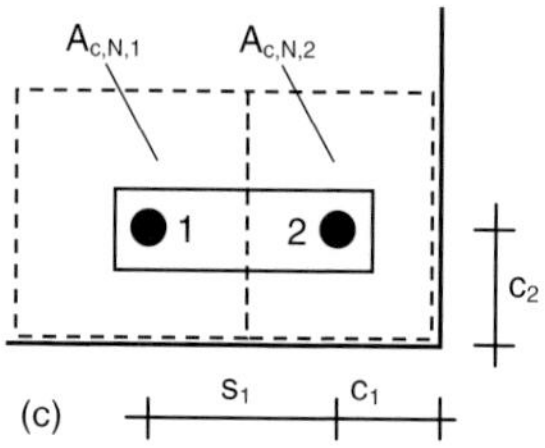

(c)

$$A_{c,N,1} = (0.5 \cdot s_{cr,N} + 0.5 \cdot s_1) \cdot (0.5 \cdot s_{cr,N} + c_2)$$
$$A_{c,N,2} = (0.5 \cdot s_1 + c_1) \cdot (0.5 \cdot s_{cr,N} + c_2)$$
$$s_1 \leq s_{cr,N};\ c_1,\ c_2 \leq c_{cr,N}$$

Fig. 5.19 Examples of the calculation of the base area $A_{c,N}$ in case of concrete pry-out failure and verification of the most unfavourable individual fastener of a group loaded by a torsion moment
(a) quadruple fastening without edge influence
(b) fastening with six fasteners without edge influence
(c) double fastening in the component corner

5.3.5 Concrete edge failure

Anchorages close to an edge subjected to a shear load perpendicular to the edge may fail via fracture of the concrete. On the concrete surface the fracture crack presents an angle with respect to the edge of about 35 ° on average. Then, the length of the concrete break-out body is approximately 3 times the value of the edge distance c_1 of the fastener (Figure 5.20a). The depth at the face of the edge is assumed to about 1.5 times the edge distance $(1{,}5 \cdot c_1)$ with sufficient accuracy.

For fastenings with more than one edge, the resistances for all edges shall be calculated if a component of the shear load acts parallel to the edge or in the direction of the edge under consideration. Shear forces acting away from the edge may be neglected and the verification of concrete edge failure is not necessary. However, this verification is substituted by verification of the failure mode pry-out.

The characteristic concrete edge resistance of group fastenings depends on the concrete strength, the behaviour of the concrete in tension and the condition of the base material

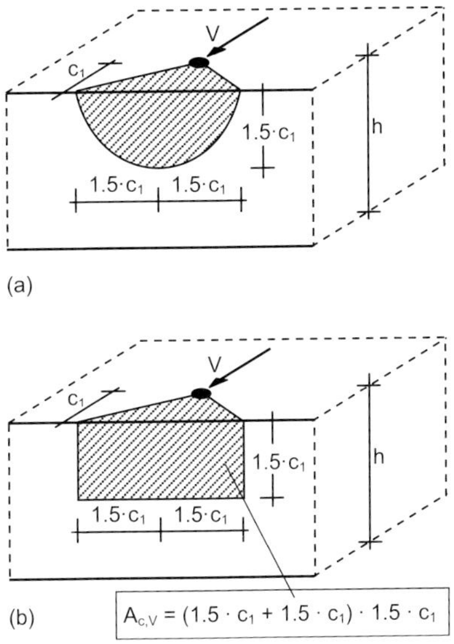

Fig. 5.20 Concrete edge break-out body of a single fastener under shear load (Fuchs, Eligehausen, and Breen (1995))
(a) schematically
(b) projected surface $A_{c,V}$ of the idealized break-out body

(cracked or uncracked) as well as the edge distance and the stiffness of the fasteners. Besides spacing to adjacent fasteners as well as the thickness of the structural member are of importance. In addition, the eccentricities of the load acting on the fasteners within a group, the load direction with respect to the edge of the component and the edge reinforcement have to be considered. The characteristic resistance of a fastener group corresponds to:

$$V_{Rk,c} = V^0_{Rk,c} \cdot \frac{A_{c,V}}{A^0_{c,V}} \cdot \psi_{s,V} \cdot \psi_{h,V} \cdot \psi_{ec,V} \cdot \psi_{\alpha,V} \cdot \psi_{re,V} \tag{5.7}$$

The different factors of Equation 5.7 are given below.

5.3.5.1 Characteristic resistance of a single fastener

The characteristic resistance of a post-installed fastener loaded perpendicular to the edge in cracked concrete according to Hofmann (2004) corresponds to:

$$V^0_{Rk,c} = 1.6 \cdot d^{\alpha}_{nom} \cdot l^{\beta}_{f} \cdot \sqrt{f_{ck,cube}} \cdot c_1^{1,5} \quad [\text{N}] \tag{5.7a}$$

with:

d_{nom} outer diameter of the fastener, see ETA [mm] ≤ 60 mm
l_f effective length of the fastener, see ETA [mm]
$= h_{ef}$ in case of a uniform diameter of the fastener $\leq 8\ d_{nom}$

$$a = 0.1 \cdot \left(\frac{l_f}{c_1}\right)^{0.5} \tag{5.7a$_1$}$$

$$\beta = 0.1 \cdot \left(\frac{d_{nom}}{c_1}\right)^{0.2} \tag{5.7a$_2$}$$

$f_{ck,cube}$ characteristic concrete compressive strength [N/mm^2] for cubes with a side length of 150 mm considering the limitations given in the relevant ETA

c_1 edge distance in the direction of the shear load [mm]

If post-installed fasteners with $d_{nom} > 60$ mm or $l_f > 8 \cdot d_{nom}$ shall be designed, then the limiting values of $d_{nom} = 60$ mm or $l_f = 8 \cdot d_{nom}$ have to be inserted in Equations 5.7a, 5.7a$_1$ and 5.7a$_2$.

Equation 5.7a is valid for applications in cracked concrete. The factor $\psi_{re,V}$ (Section 5.3.5.7) takes account of a possible crack effect.

5.3.5.2 Effect of spacing

The term $A_{c,V}/A^0_{c,V}$ in Equation 5.7 takes into account the influence of fastener spacing on the characteristic concrete edge break-out resistance. Figure 5.20a shows that the length of the break-out body of a single fastener close to the edge corresponds to 3 times the edge distance c_1. In case of fasteners with a spacing $s \geq 3 \cdot c_1$ within a group, the break-out bodies of adjacent fasteners do not overlap and the characteristic resistance of the fastener group equals n times the resistance of a single fastener having the same edge distance (n = number of the fasteners close to the edge). Anchor groups with spacing $s < 3c_1$ form a common concrete fracture surface since the concrete break-out bodies overlap. Then the fractured surface available for the load introduction into the concrete becomes smaller than the sum of the fractured surfaces of the same number of individual fasteners. This results in a reduction of the characteristic resistance.

The effect of fastener spacing on the characteristic resistance of a group of fasteners is taken into account in the verification in a manner directly analogous to the tension loading case with concrete cone failure by comparing the fractured surface of the fastener group projected to the side face of the component with the sum of the fractured surfaces of the same number of individual fasteners. However, the calculation of overlapping surfaces due to the curved shape (Figure 5.20a) is rather complex. Therefore, the approach described in CEN/TS assumes idealized break-out bodies. It is based on a proposal of Fuchs (1991) and Fuchs, Eligehausen, and Breen (1995).

The concrete break-out body is idealized by a half pyramid with the height c_1 and the base length of $3 \cdot c_1$ and $1.5 \cdot c_1$ (Figure 5.20b). The influence of spacing on the characteristic resistance of a post-installed fastener group is calculated by comparing the projected surface of the idealized break-out body of the group ($A_{c,V}$) with the value of a single post-installed fastener ($A^0_{c,V}$). Figure 5.21 shows examples of different post-installed fastener groups. In Figure 5.22 the method is explained with help of examples. Figure 5.22a shows a double fastening in the corner of a component with edge distances c_1 = to 100 mm and $c_2 = 80$ mm. The spacing $s = 200$ mm is smaller than $3\ c_1 = 300$ mm. Hence, a common break-out body develops. The edge distance parallel to the load direction load c_2 is

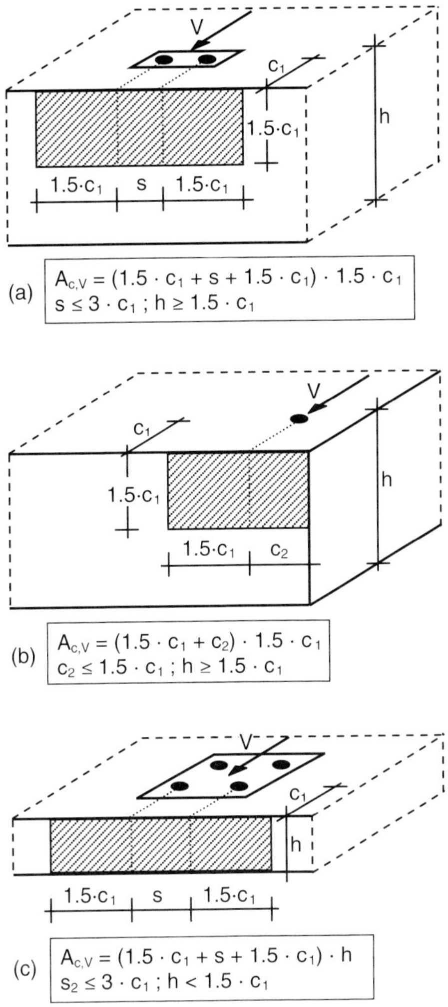

Fig. 5.21 Definition of the projected area of the idealized rupture body of different fastenings with concrete edge break-out
(a) double fastening on the edge of a thick component
(b) single fastener in the corner of a thick component
(c) quadruple fastening in the edge of a thin component

smaller than 1.5-times the edge distance c_1, i.e., the break-out body is not able to form completely to this side. Figure 5.22b shows the same application, indeed, the spacing $s = 320$ mm is bigger than $3\ c_1$. Therefore, the break-out bodies of both post-installed fasteners do not overlap, the spacing has no influence and the base plate is fixed by two individual post-installed fasteners independent of each other.

If the thickness of the concrete member is $h < 1.5\ c_1$, the break-out body is intersected by the lower edge of the component and cannot develop completely.

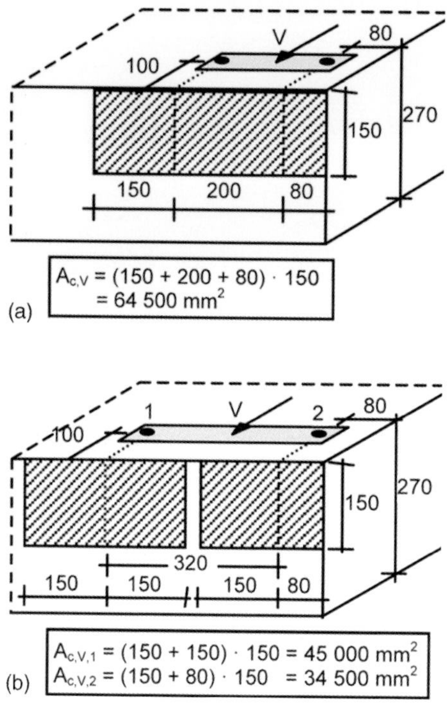

Fig. 5.22 Determination of the projected area of the idealized break-out body at the example of a double fastening on the component edge (edge distance $c_1 = 100$ mm)
(a) spacing $s_1 = 200\,\text{mm} < 3\,c_1$
(b) spacing $s_1 = 320\,\text{mm} > 3\,c_1$

This yields a decrease of the characteristic resistance and is taken into account in the calculation of the projected area (Figure 5.21c).

If groups with post-installed fasteners are arranged perpendicular to the edge and are loaded by an edge-parallel shear force or a torsion moment, and the spacing s_l perpendicular to the edge is smaller than the edge distance c_1 and $c_1 < 150$ mm, then the described approach can lead to the results which are unsafe. This is explained in the following by means of a group with two post-installed fasteners which is loaded by an edge-parallel shear load (Figure 5.23). By definition a load-parallel shear force is distributed to all fasteners of the group (Figure 5.23a). This means that the verification of the post-installed fastener close to edge is performed with the shear load $V_{Ed}/2$. If the spacing s_1 is relatively small, however, the post-installed fastener in the rear activates only a slightly bigger fractured surface than the fastener in front. This means that the crack of the rear post-installed fastener – as shown Figure 5.23a – does not propagate in parallel with the crack initiated by the front fastener but – in search of the shortest way to reach an energy minimum – connects to the front crack (Figure 5.23b). Therefore, the resistance of the rear post-installed fastener for concrete edge failure exceeds only slightly the resistance of the post-installed fastener close to edge. A conservative

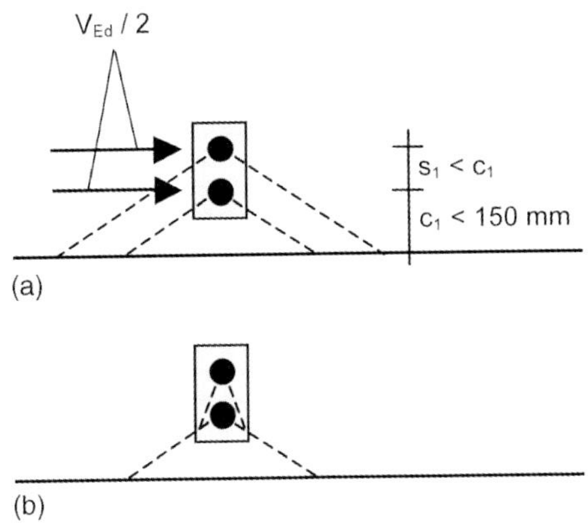

Fig. 5.23 Double fastening vertically to the component edge with low spacing and edge distance, loaded by shear load acting parallel to the edge
(a) cracks of both fasteners develop in parallel
(b) cracks of the rear fastener intersects the cracks of the fastener close to edge

approach would be to verify the front post-installed fastener for the entire shear load V_{Ed} in cases as shown in Figure 5.23. Then, however, the equilibrium conditions would be disregarded. CEN/TS provides no information how to proceed in such cases. Here clarification is needed.

5.3.5.3 Effect of edge distances parallel to the load direction

With applications in the component corner (edge distance $c_2 \leq 1.5\ c_1$) or in a narrow component (two edges with distances $c_{2,1}, c_{2,2} \leq 1.5\ c_1$) an additional edge influence is to be considered. The geometrical influence of the reduction of the fractured surfaces by load-parallel edges is taken into account by the projected area $A_{c,V}$ (Figure 5.21b). In addition, load-parallel edges disturb the condition of stress in the concrete. This is considered by the $\psi_{s,V}$ factor after Equation 5.7b.

$$\psi_{s,V} = 0,7 + 0,3 \cdot \frac{c_2}{1.5 \cdot c_1} \leq 1 \tag{5.7b}$$

In case of two load-parallel edges (narrow component) the smaller of both edge distances $c_{2,1}$ and $c_{2,2}$ is inserted in Equation 5.7b.

5.3.5.4 Effect of member thickness

If a component is thicker than 1.5-times the edge distance c_1, the concrete break-out body on the side of the component can form completely. If this is not the case, the break-out body is cut off by the lower edge of the component and the fracture surface available to the introduction of the shear load is reduced. This influence is already considered by the surface $A_{c,V}$ (Figure 5.21c). Then the characteristic resistance for concrete edge break-out is proportional to the component thickness for $h < 1.5\ c_1$. Tests show that this approach is conservative because the load-carrying capacity of fastening in reality does not decrease proportionally to the component thickness h but is reduced less (Zhao, Fuchs, and Eligehausen (1989)). This is considered in Equation 5.7c by the factor $\psi_{h,V}$.

$$\psi_{h,V} = \left(\frac{1.5 \cdot c_1}{h}\right)^{0.5} \geq 1 \tag{5.7c}$$

It is to be noted that $\psi_{h,V}$ is an *enlargement factor*. This factor must be always greater than 1 because it compensates the linear dependence of the characteristic resistance assumed with the projected area $A_{c,V}$. If the factor $\psi_{h,V}$ is smaller than 1 for thick components ($h > 1.5\ c_1$), the value may be assumed to 1.

5.3.5.5 Effect of the eccentricity of the load

If a shear force does not act in the centre of gravity of the fastener group and/or a torsion moment T_{Ed} is applied, in addition, the individual post-installed fasteners have to resist different shear loads. This is considered by the factor $\psi_{ec,V}$, see Equation 5.7d.

$$\psi_{ec,V} = \frac{1}{1 + 2 \cdot e_V/(3 \cdot c_1)} \leq 1 \tag{5.7d}$$

When calculating the eccentricity e_v it is to be noted that the shear load components which are directed reverse from the edge may be neglected in the verification of concrete edge failure (see Section 4.3.1 and Figure 4.11). Figure 5.24 demonstrates the determination of e_v.

The double fastening in Figure 5.24a is loaded by a torsion moment. The torsion moment causes that only the left post-installed fastener is loaded in the direction to the edge. By definition the component directed reverse from the edge (post-installed fasteners on the right) may be neglected with the verification of concrete edge breakout. Herewith the eccentricity of the shear load corresponding to the centre of gravity of the fastener group is to $e_V = s/2$.

The double fastening in Figure 5.24b is loaded by an inclined shear force and a torsion moment where the torsion portion dominates. As described in Section 4.3.1, the inclined shear force is decomposed in a horizontal and a vertical component and the forces are assigned to both post-installed fasteners equally. For double fastenings the shares resulting from the torsion moment are established by dividing the torsion moment by the spacing s. For the post-installed fastener on the right the torsion component is bigger than the vertical portion from the shear load, the sum of both forces is directed reverse from the edge and may be neglected, therefore. The post-installed fastener group is loaded by an inclined shear force whose resultant acts on the post-installed fastener on the left. The eccentricity e_V results from the length of the perpendicular of the centre of gravity of the post-installed fastener group on the load.

In the example of Figure 5.24c the vertical load portion directed to the edge resulting from the shear force is bigger than the value calculated from torsion. Therefore, both post-installed fasteners are loaded in the direction of the component edge. First the resultant of the vertical components and their position is calculated (in the example on

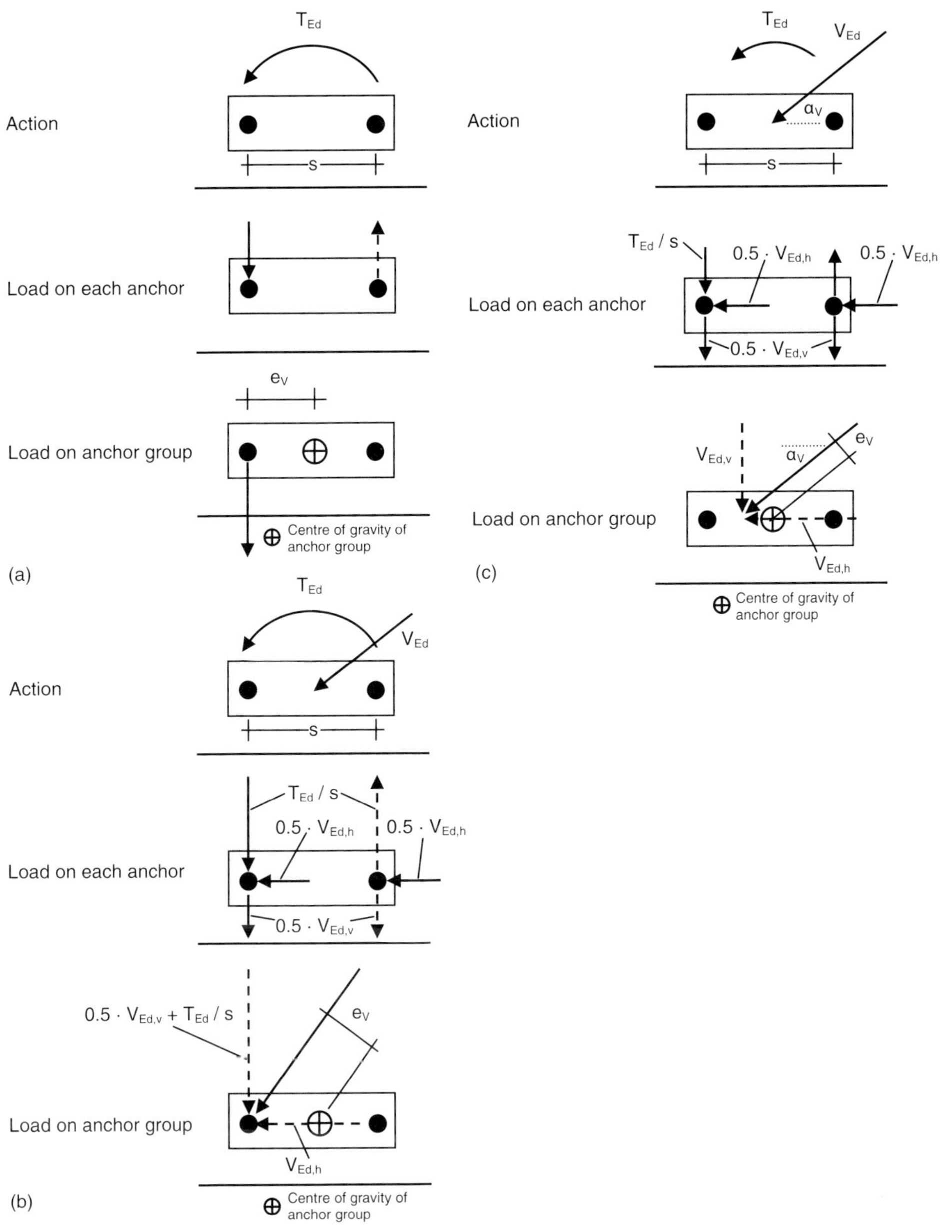

Fig. 5.24 Examples of the determination of the eccentricity e_v for the verification of concrete edge failure

(a) double fastening, loaded by a torsion moment

(b) double fastening, loaded by a shear force acting inclined to the edge and a torsion moment (influence of the torsion greater than that of the shear force)

(c) double fastening, loaded by a shear force acting inclined to the edge and a torsion moment (influence of the torsion smaller than that to the shear force)

the left from the centre of gravity of both post-installed fasteners). Then the value and the load angle of the resultant are determined by superimposition with the horizontal shear load components. Since the vertical shares of the shear force from the torsion moment eliminate, the resultant corresponds to the shear force V_{Ed}. Then the eccentricity e_v is calculated again from the length of the perpendicular of the centre of gravity of the post-installed fastener group on this resultant.

5.3.5.6 Effect of load direction

The load direction is considered by the factor $\psi_{\alpha,V}$, see Equation 5.7e

$$\psi_{\alpha,V} = \sqrt{\frac{1}{(\cos\alpha_V)^2 + (0,4 \cdot \sin\alpha_V)^2}} \geq 1 \tag{5.7e}$$

α_V is defined as inclination between the resulting shear force of the fasteners close to the edge under consideration and a line perpendicular to the edge (Figure 5.25). This angle is not necessarily identical with the angle of the external shear force acting on the fastening. Examples show the Figures 5.24b and 5.26. If the quadruple fastening is loaded by a shear force inclined to the component edge (Figure 5.26a), the load component acting perpendicular to the edge is resisted only by both post-installed fasteners close to edge, while the edge-parallel component is distributed to the fasteners far from and close to edge (Figure 5.26b) for abovementioned reasons. Therefore the angle α_V of the resultant shear load acting on the post-installed fasteners close to edge is smaller than the inclination of the external shear force.

As already mentioned, the verification of concrete edge failure must be performed for all component edges. Figure 5.27 demonstrates this rule by means of a fastening in the component corner.

Figure 5.27a shows the application with the relevant dimensions, forces and values specific for the post-installed fasteners. In Figures 5.27b and c the verification of concrete edge failure of the edges 1 (horizontal edge) and 2 (vertical edge) is shown. The shear loads acting on the governing post-installed fasteners close to the edge are calculated according to Figure 4.10. From this result the inclinations $\alpha_V = 33.7°$ and $\alpha_V = 20.6°$ (measured between the resultant shear load and the perpendicular on the

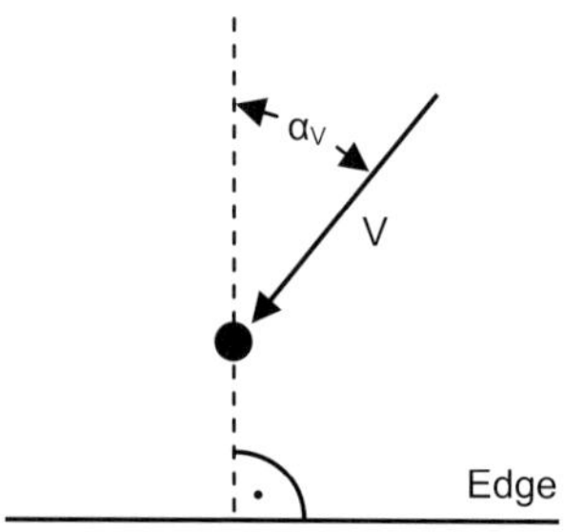

Fig. 5.25 Definition of the angle α_V for the verification of concrete edge failure

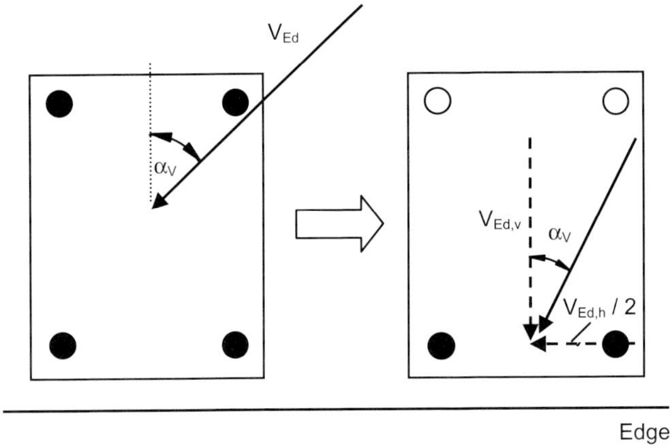

Fig. 5.26 Quadruple fastening in the component edge, loaded by an oblique shear force

edge to be verified). The calculation indicates that the vertical edge 2 governs for concrete edge failure.

5.3.5.7 Effect of the position of the fastening

The factor $\psi_{re,V}$ in Equation 5.7 takes account of the effect of the position of the fastening in cracked or uncracked concrete and of the positive influence of edge reinforcement on the resistance for concrete edge failure.

$\psi_{re,V} = 1.0$ fastening in cracked concrete without edge reinforcement or stirrups
$\psi_{re,V} = 1.2$ fastening in cracked concrete with straight edge reinforcement ($\geq$ Ø 12 mm)
$\psi_{re,V} = 1.4$ fastening in cracked concrete with straight edge reinforcement ($\geq$ Ø 12 mm) and closely spaced stirrups or wire mesh with a spacing $a \leq 100$ mm and $a \leq 2 \cdot c_1$ or fastening in uncracked concrete

By the limitation of the spacing of the reinforcing bars (stirrups or bars of wire mesh) it is ensured that the reinforcement is sufficiently anchored in the concrete break-out body and therefore is able to work as supplementary reinforcement.

5.3.5.8 Special case: narrow thin member

As in case of cone shaped concrete break-out (see Section 5.2.4.5) it exists for the verification of concrete edge break-out one application where Equation 5.7 leads to conservative results. This is valid for fastenings in a narrow and thin concrete member where $c_{2,1}$ and $c_{2,2}$ as well as the member thickness h are smaller than 1.5-times the edge distance c_1 in the direction of the load. This means that the concrete break-out body can form neither to the lateral edges nor to the below end of the component completely. Figure 5.28 presents the facts by means of an example.

Figure 5.28 shows a double fastening ($s = 200$ mm) at the end of a narrow thin component ($h = 150$ mm). Cracked concrete C20/25 and a straight edge reinforcement

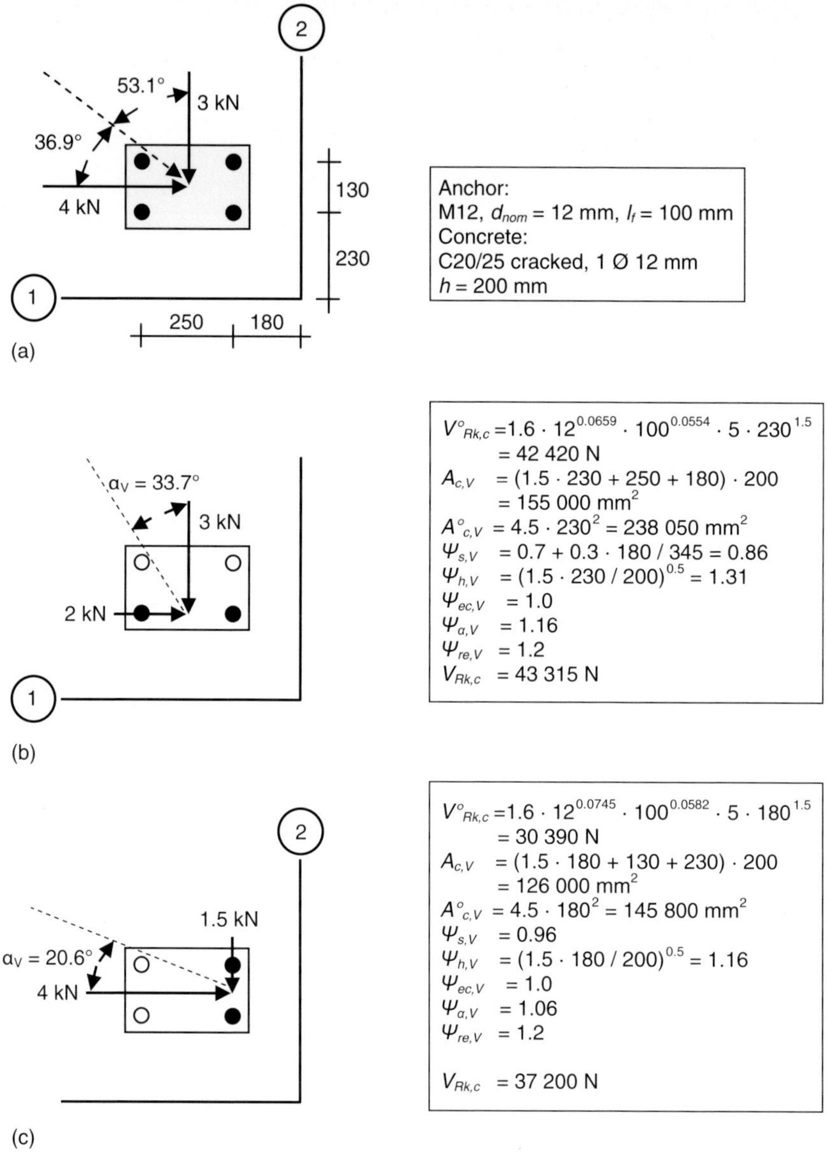

Fig. 5.27 Required verifications in case of concrete edge failure for a quadruple fastening in the component corner

with stirrups (a = 100 mm) are assumed. The edge distance in load direction is in the first case $c_1 = 120$ mm and in second $c_1 = 240$ mm. In both cases the edge distances to the lateral edges are $c_{2,1} = 130$ mm and $c_{2,2} = 100$ mm. The fastening consist of post-installed fasteners of size M12 ($d_{nom} = 12$ mm) and with an embedment depth $h_{ef} = 80$ mm ($l_f = 80$ mm). After calculation according to Equation 5.7 the following is obtained:

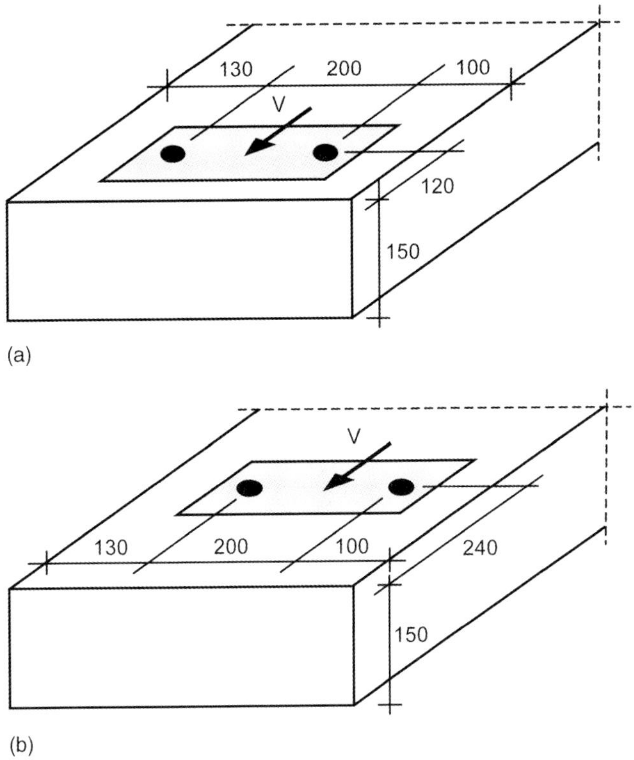

Fig. 5.28 Fastening in a narrow thin component
(a) edge distance $c_1 = 120\,\text{mm}$
(b) edge distance $c_1 = 240\,\text{mm}$

Case 1:

$$V_{Rk,c} = (1.6 \cdot 12^{0.0816} \cdot 80^{0.0631} \cdot \sqrt{25} \cdot 120^{1.5}) \cdot \frac{64\,500}{64\,800} \cdot 0.867 \cdot 1.095 \cdot 1 \cdot 1 \cdot 1{,}4$$
$$= 22\,470\,\text{N}$$

Case 2:

$$V_{Rk,c} = (1.6 \cdot 12^{0.0577} \cdot 80^{0.0549} \cdot \sqrt{25} \cdot 240^{1.5}) \cdot \frac{64\,500}{259\,200} \cdot 0.783 \cdot 1.549 \cdot 1 \cdot 1 \cdot 1{,}4$$
$$= 18\,450\,\text{N}$$

The resistance in the second case in spite of the bigger edge distance c_1 is smaller than in case 1. In reality the same resistance would be obtained in both cases, because the narrow and thin component will fracture always at the position of the fastening independent of the edge distance. More exact results are yielded if the modified distance c_1' is used instead of the actual edge distance c_1.

$$c_1' = \max\{c_{2,max}/1.5;\quad h/1.5;\quad s_{max}/3\} \tag{5.7f}$$

with $c_{2,\mathrm{max}}$ as the largest of the two edge distances parallel to the direction of loading. Based on this assumption for the example given in Figure 5.28 is:

$$c_{2,max}/1.5 = 130/1.5 = 86,7\,\mathrm{mm}$$

$$h/1.5 = 150/1.5 = 100\,\mathrm{mm}$$

$$s_{max}/3 = 200/3 = 66.67\,\mathrm{mm}$$

The largest of the three values above is governing. Therefore $c_1' = 100\,\mathrm{mm}$ and the same characteristic resistance is obtained in both cases:

$$\begin{aligned} V_{Rk,c} &= (1.6 \cdot 12^{0.0894} \cdot 80^{0.0654} \cdot \sqrt{25} \cdot 100^{1.5}) \cdot \frac{64\,500}{45\,000} \cdot 0.90 \cdot 1 \cdot 1 \cdot 1 \cdot 1.4 \\ &= 24\ 030\,\mathrm{N} \end{aligned}$$

The calculated characteristic resistance corresponds to the maximum value of the assumed geometry. By reducing the edge distance c_1 the characteristic resistance decreases and by increasing the edge distance it remains constant.

The described weak point of the design method can be fixed by the introduction of the modified edge distance c_1'.

5.4 Combined tension and shear load

Fastenings simultaneously subjected to tension and shear loading shall be verified according to Sections 5.2 and 5.3 and, in addition for combined tension and shear load. Two cases are distinguished.

5.4.1 Steel failure decisive for tension and shear load

In case of governing steel failure for the verification in both, the load directions tension and shear Equation 5.8 shall be satisfied.

$$\beta_N^2 + \beta_V^2 \leq 1 \tag{5.8a}$$

$$\beta_N = N_{Ed}/N_{Rd} \leq 1 \tag{5.8b}$$

$$\beta_V = V_{Ed}/V_{Rd} \leq 1 \tag{5.8c}$$

5.4.2 Other modes of failure decisive

If concrete failure is governing at least one of both verified load directions (tension load: pull out/pull through, conical concrete break-out or splitting; shear load: concrete pry-out or concrete edge break-out), then both Equations 5.9 or 5.10 are valid:

$$\beta_N^{1.5} + \beta_V^{1.5} \leq 1 \tag{5.9a}$$

$$\beta_N = N_{Ed}/N_{Rd} \leq 1 \tag{5.9b}$$

$$\beta_V = V_{Ed}/V_{Rd} \leq 1 \tag{5.9c}$$

$$\beta_N + \beta_V \leq 1.2 \tag{5.10a}$$

$$\beta_N = N_{Ed}/N_{Rd} \leq 1 \tag{5.10b}$$

$$\beta_V = V_{Ed}/V_{Rd} \leq 1 \tag{5.10c}$$

The more favourable of both approaches may be used for the design. The largest value of β_N and β_V for the different failure modes is to be used in the Equations 5.9 and 5.10.

If steel failure is governing under tension and shear forces, according to CEN/TS the interaction may occur after Equation 5.8 based on a quadratic interaction. This verification is not conservative if the design values for concrete failure are only slightly below the values for steel failure. Hence, either the proof should occur after Equations 5.9 or 5.10 where steel failure is to be observed for the determination of β_N and β_V, or the improved approved approach of fib (2011) should be used. This approach was not available when CEN/TS developed. Then steel failure and concrete failure are always verified and both verifications must be satisfied and, hence, the more unfavourable case governs the design. This approach should be considered with in the upcoming revision of CEN/TS.

6 Verification of post-installed fasteners (chemical systems) for the ultimate limit state based on the theory of elasticity

6.1 General

The design rules of CEN/TS, Part 5 described in the following sections are valid for post-installed adhesive fasteners as shown in Figure 2.1e_1, consisting of a threaded rod or internally threaded sleeve and mortar. The mortar is provided either in glass cartridges, plastic cartridges or foil bags and is mixed by driving the threaded rod or internal thread sleeve (rotary/hammer mode) or, however, is pre-measured in cartridges and mixed in special mixing nozzles while injecting into the borehole. The load introduction to the concrete occurs via bond along the entire embedment depth. On the other hand the working principle of bonded expansion anchors according to Figure 2.1e_2 corresponds to the function of torque-controlled mechanical fasteners (Figures 2.1a_1 and 2.1a_2). In case of bonded expansion anchors by application of an installation torque the fully cured mortar is split in single segments of mortar which in principle work like the segments of the sleeve of a torque-controlled expansion fastener. Therefore, bonded expansion anchors are not designed according to the rules described in this section, but as torque-controlled mechanical of the sleeve and bolt type according to Section 5 or CEN/TS, Part 4, respectively.

While CEN/TS, Part 4 for post-installed mechanical fastening systems includes three design procedures (see Section 5.1) which clearly differ in their complexity and in the conservatism of the calculated results, only one procedure is given in CEN/TS, Part 5, for post-installed chemical systems. It corresponds in the essentials to the design procedure *A* for mechanical systems, offers solutions for almost any desired applications and leads to the best results.

The design according to CEN/TS, Part 5 is based on the rules for execution as described in Section 5.1 concerning the compaction of the concrete, the production of the boreholes, the avoidance of the damage of the reinforcement including the necessary distances to tendons, the cleaning of the boreholes as well as the treatment of false drilled holes. The examination and acceptance of the installation is to be carried out – as for post-installed mechanical fastening systems – by qualified personnel.

The load-carrying capacity of adhesive fasteners is influenced significantly by the quality of the installation. Hence, the installation is to be carried out by trained installers after the installation instructions of the manufacturer under use of the required tools. The careful cleaning of the boreholes is especially important.

6.2 Tension load

6.2.1 Required verifications

In principle the same failure modes are observed with post-installed chemical fasteners as for post-installed mechanical fasteners (see Section 5.2.1). Chemical fasteners introduce the load into the base material continuously along the entire embedment

Design of Fastenings for Use in Concrete – the CEN/TS 1992-4 Provisions. First edition.
Rainer Mallée, Werner Fuchs, Rolf Eligehausen.

Fig. 6.1 Pulled out chemical fastener with combined pull-out and concrete break-out after a test (Eligehausen, Mallée, and Rehm (1984))

depth. Therefore the failure mode pull-out/pull-through as observed for post-installed mechanical fasteners does not occur in case of bonded fasteners. In lieu of these failure modes bonded anchors might fail by combined pull-out and concrete break-out. Other than with expansion fasteners and undercut fasteners the break-out cone does not propagate from the end of the fasteners but dependent on the product forms from 0.3 to 0.7-times the value of the embedment length (Figure 6.1). On the remaining length of the threaded rod the bond between mortar and concrete or between threaded rod and mortar is destroyed. Often a mixed bond failure appears (in the upper part the bond between mortar and concrete and in the lower part between threaded rod and mortar is destroyed). Figure 6.2 shows the possible modes of bond failures (after Cook *et al.* (1998)).

6.2.2 Steel failure

In case of steel failure the threaded rod or the internally threaded sleeve is ruptured. Section 5.2.2 applies without change.

6.2.3 Combined pull-out and concrete failure

Combined pull-out and concrete break-out appears only if the characteristic bond strength τ_{Rk} of bonded fastener according to ETA is smaller than the value $\tau_{Rk,max}$. This value corresponds to the bond strength which is necessary to generate a concrete break-out developing from the end of the threaded rod. $\tau_{Rk,max}$ is calculated by equating the concrete break-out load $N^0_{Rk,c}$ after Equation 5.2a

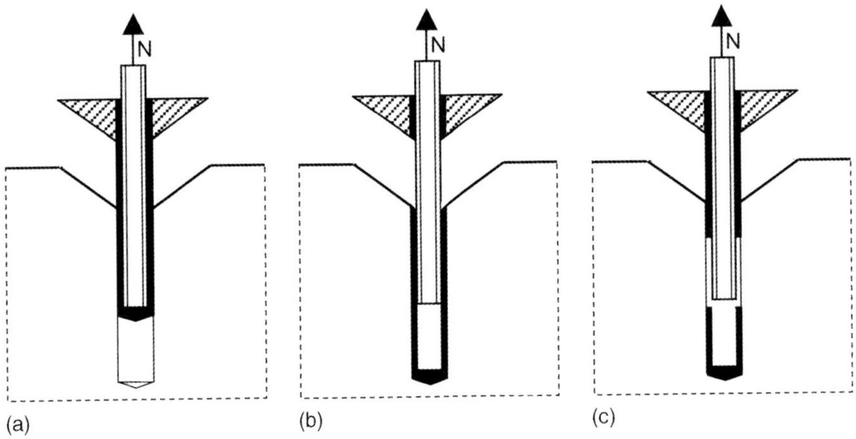

Fig. 6.2 Failure modes of bonded fasteners (Cook *et al.* (1998))
(a) failure in the interface between mortar and bore wall
(b) failure in the interface between mortar and threaded rod
(c) combined failure

and the basic value $N^0_{Rk,p}$ for bond failure after Equation 6.2a. The value $\tau_{Rk,max}$ is calculated as follows:

$$\tau_{Rk,\max} = \frac{k_8}{\pi \cdot d} \cdot \sqrt{h_{ef} \cdot f_{c,cube}} \tag{6.1}$$

mit:

k_8 given in the ETA, factor for cracked and non-cracked concrete
d thread diameter [mm]
h_{ef} embedment depth [mm]
$f_{ck,cube}$ characteristic concrete compressive strength [N/mm²] for cubes with a side length of 150 mm considering the limitations given in the relevant ETA

According to current experience the value for k_8 is 7.2 for applications in cracked concrete and $k_8 = 10.1$ for applications in non-cracked concrete.

The characteristic resistance of a fastener group in case of combined pull-out and concrete break-out is influenced by numerous parameters, for example, by the bond strength, the state of the concrete (cracked or uncracked), the embedment depth, by spacing to neighbouring fasteners of a fastener group as well as distances to free component edges, by a potential eccentricity of the load acting on the fasteners within a group as well as by an unfavorably effective heavy surface reinforcement. The characteristic resistance can be calculated taking into account these influencing parameters according to Equation 6.2 (Eligehausen *et al.* (2005)), Eligehausen, Cook, and Appl (2006) and Appl (2009)):

$$N_{Rk,p} = N^0_{Rk,p} \cdot \frac{A_{p,N}}{A^0_{p,N}} \cdot \psi_{s,Np} \cdot \psi_{g,Np} \cdot \psi_{re,N} \cdot \psi_{ec,Np} \quad [\mathrm{N}] \tag{6.2}$$

The individual factors of Equation 6.2 are explained in the following.

6.2.3.1 Characteristic resistance of a single fastener

The characteristic resistance of a single post-installed fastener is calculated under the assumption of a bond strength uniformly distributed along the embedment depth (linear distribution of the steel stress) and therefore results from the product of the characteristic bond strength τ_{Rk} and the surface area:

$$N^0_{Rk,p} = \tau_{RK} \cdot \pi \cdot d \cdot h_{ef} \tag{6.2a}$$

with:

τ_{Rk} characteristic bond resistance, depending on the concrete strength class, $\tau_{Rk,cr}$ for cracked concrete or $\tau_{Rk,ucr}$ uncracked concrete are given in the corresponding ETA

It is obvious that the assumption of uniform bond strength cannot be valid for arbitrarily big embedment depths. Therefore, CEN/TS, Part 5 limits the application to bonded fasteners with $h_{ef} \leq 20 \cdot d$. Post-installed chemical fasteners with bigger anchorage depths are not covered currently. The limiting value $h_{ef} = 20 \cdot d$ is rather conservative after Cook *et al.* (1998)). The authors assume a uniform bond stress distribution up to an embedment depth $h_{ef} = 25 \cdot d$. On the other hand Meszaros (1999) sets $h_{ef} = 20 \cdot d$ as limiting value of the embedment depth for a uniform bond strength.

The characteristic bond strength τ_{Rk} is given in ETA. Besides, depending on the approval it is distinguished between applications in the cracked and uncracked concrete. Moreover, the approval can contain values of different intensities of bore cleaning, dry or wet concrete, water-filled boreholes as well as the application in different temperature ranges. The respective characteristic strengths are determined based on the results of extensive prequalification testing according to European Organization for Technical Approvals (EOTA) (2008).

6.2.3.2 Edge distance and spacing

Spacing

The quotient $A_{p,N}/A^0_{p,N}$ in Equation 6.2 considers the effect of spacing and edge distances on the characteristic resistance for combined pull-out and concrete break-out failure. The areas $A_{p,N}$ and $A^0_{p,N}$ are determined such as the projected areas $A_{c,N}$ and $A^0_{c,N}$ in case of concrete cone failure. However, the characteristic spacing and edge distance $s_{cr,N}$ and $c_{cr,N}$ have to be replaced by the values $s_{cr,Np}$ and $c_{cr,Np}$. Examples for the calculation of the projected areas are given in Figures 5.5–5.7.

Characteristic spacing $s_{cr,Np}$ and characteristic edge distance $c_{cr,Np}$ are calculated with Equation 6.2b:

$$s_{cr,Np} = 7.3 \cdot d \cdot \sqrt{\tau_{Rk}} \leq 3 \cdot h_{ef} \quad (6.2b1)$$

$$c_{cr,Np} = s_{cr,Np}/2 \quad (6.2b2)$$

with:

d thread diameter [mm]

τ_{Rk} characteristic bond strength [N/mm^2] for *uncracked* concrete C20/25

Edge distances

As already shown in case of concrete cone failure the geometric effect of component edges on the characteristic resistance for combined pull-out and concrete break-out can also be determined by the quotient $A_{p,N}/A^0_{p,N}$. If a fastening with an edge distance $c < c_{cr,\ Np}$ is arranged, the break-out body overlaps with the component edge and the available fractured surface is reduced. Examples are shown in Figures 5.9 and 5.10. Besides, the characteristic spacing and edge distance $s_{cr,\ N}$ and $c_{cr,\ N}$ have to be substituted again with the values $s_{cr,Np}$ and $c_{cr,Np}$.

In addition to the described geometrical influence of edges the disturbance of the rotation-symmetrical stress condition by component edges is also to be considered in case of combined pull-out and concrete break-out (Figure 5.11). This effect is taken into account in Equation 6.2 by the factor $\psi_{s,Np}$.

$$\psi_{s,Np} = 0.7 + 0.3 \cdot \frac{c}{c_{cr,Np}} \leq 1,0 \quad (6.2c)$$

For fastenings with several edge distances (e.g. fastening in a corner of the concrete member or in a narrow member), the smallest edge distance c shall be inserted in Equation 6.2c.

6.2.3.3 Effect of closely spaced fasteners

Equation 6.2 comprises the so-called group factor $\psi_{g,Np}$. This factor considers the influence of the surface of the break-out body for fastener groups (Figure 6.3). For double fastenings with a spacing $s = d$, the fractured surface in case of bond failure and thus the characteristic resistance of the double fastening is bigger by the factor $\sqrt{n}$ than the surface corresponding to a single chemical fastener (Figures 6.3a$_1$ and 6.3a$_2$). This is considered by the group factor $\psi_{g,Np}$ (Eligehausen *et al.* (2005)):

$$\psi_{g,Np} = \psi^0_{g,Np} - \sqrt{\frac{s}{s_{cr,Np}}} \cdot \left(\psi^0_{g,Np} - 1\right) \geq 1 \quad (6.2d)$$

with:

$$\psi^0_{g,Np} = \sqrt{n} - \left(\sqrt{n} - 1\right) \cdot \left(\frac{\tau_{Rk}}{\tau_{Rk,\max}}\right)^{1,5} \geq 1 \quad (6.2e)$$

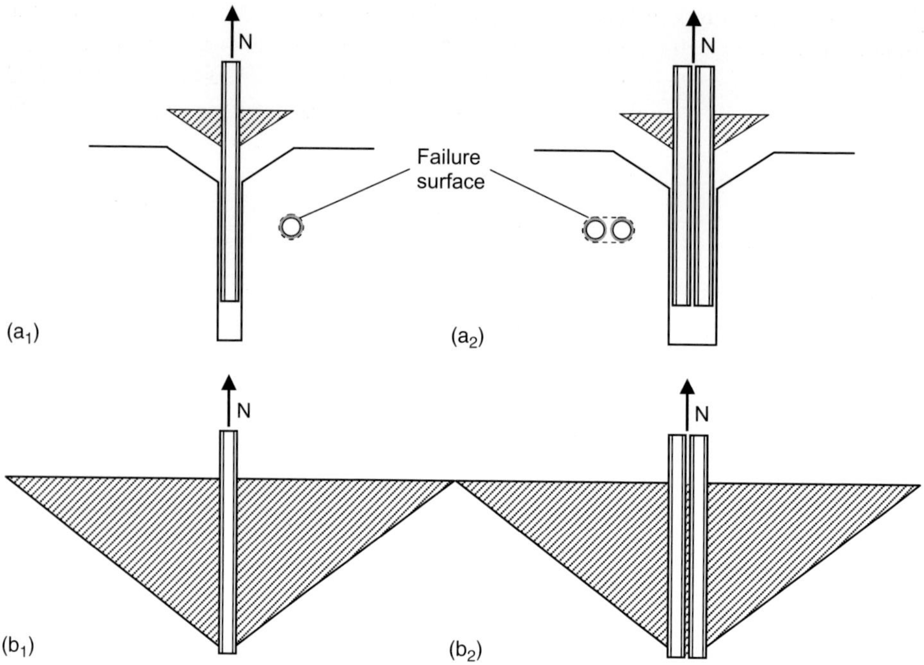

Fig. 6.3 Failure of a double fastening with bonded fasteners arranged with low spacing (Eligehausen *et al.* (2005))
(a) combined pull-out and concrete break-out
(b) conical concrete break-out

n	number of fasteners in a group
τ_{Rk}	characteristic bond resistance, depending on the concrete strength class, $\tau_{Rk,cr}$ for cracked concrete or $\tau_{Rk,ucr}$ for uncracked concrete are given in the corresponding ETA
$\tau_{Rk,max}$	see Equation 6.1
s	spacing [mm], in case of multiple spacing the mean value of spacing shall be used
$s_{cr,Np}$	see Equation 6.2b$_1$

The group factor decrease from $\psi_{g,Np} = \psi^0_{g,Np}$ for $s = 0$ to $\psi_{g,Np} = 1$ for $s = s_{cr,Np}$.

The characteristic resistance of the double fastening in case of concrete break-out corresponds to just about the value of a single fastener (Figure 6.3), because the concrete break-out cone of the group is only slightly bigger than that of the single fastener (Figures 6.3b$_1$ and 6.3b$_2$). Hence, the factor $\psi_{g,N}$ is not required in the calculation of the characteristic resistance for this failure mode.

6.2.3.4 Effect of heavy reinforcement (shell spalling)

The load-carrying capacity in case of combined pull-out and concrete break-out failure can be adversely influenced by dense surface reinforcement – like with mechanical fasteners failing by concrete cone break-out – if chemical fasteners are anchored in the concrete cover or near the reinforcement because then the tensile stresses from the bond action of the reinforced bars overlap with those caused by the bond action of the chemical fastener. Moreover, the reinforcement might reduce the concrete area available for the transmission of tension forces ("perforating" of the concrete cover). In addition, fasteners can be anchored in an area in which the concrete strength is lower particularly in regions with dense reinforcement than in the inside cross section. Section 5.2.4.3 applies for the consideration of this negative influence consistently.

6.2.3.5 Effect of the eccentricity of the load

Section 5.2.4.4 applies analogously, indeed, in Equation 5.2d instead of the characteristic spacing $s_{cr,N}$ the value $s_{cr,Np}$ is to be inserted.

6.2.3.6 Special case: three or four edges with $c_i < c_{cr,Np}$

As discussed in Section 5.2.4.5, the CC-method leads to conservative results in applications with three or four component edges and edge distances $c_i < c_{cr,Np}$ (cf. Figure 5.13) also in the verification of combined pull-out and concrete break-out failure. Results closer to reality are achieved if the fictive value h'_{ef} is used instead of the real embedment depth h_{ef}. For the determination of the value h'_{ef} after Equations 5.2e$_1$ and 5.2e$_2$ the values $s_{cr,N}$ and $c_{cr,N}$ are to be substituted by $s_{cr,Np}$ and $c_{cr,Np}$. Further details can be taken from Section 5.2.4.5.

6.2.4 Concrete cone failure

In the verification of conical concrete break-out it is assumed that the break-out cone originates at the bottom of the threaded rod and proceeds in an angle of 35° measured towards of the horizontals up to the concrete surface (Figure 5.2). Therefore, the same conditions are given like for mechanical fasteners. Thus, Section 5.2.4 applies without changes.

6.2.5 Splitting

Section 5.2.5 is valid for the verifications of splitting caused by installation and splitting due to an external load applied to the fastener. In accordance with CEN/TS in Equation 5.3 the factor N^0_{Rk} may be replaced by the value $N^0_{Rk,c}$ after Equation 5.2a$_1$ (cracked concrete) or Equation 5.2a$_2$ (uncracked concrete). Nevertheless, the characteristic spacing and edge distances can be determined for bonded fasteners in case of $N^0_{Rk,p} < N^0_{Rk,c}$ only for bond failure. Hence, it is recommended, – like with mechanical fasteners – for chemical fasteners also to use as factor N^0_{Rk} in Equation 5.3 the minimum of the values $N^0_{Rk,c}$ after Equation 5.2a or $N^0_{Rk,p}$ after Equation 6.2a.

6.3 Shear load

6.3.1 Required verifications

Post-installed chemical fasteners subjected to shear forces exhibit the same failure modes like mechanical fasteners. Therefore, the verifications of steel failure are to be performed for shear loads without and with lever arm, for concrete pry-out and for concrete edge failure.

6.3.2 Steel failure due to shear load without and with lever arm

Sections 5.3.2 and 5.3.3 apply.

6.3.3 Concrete pry-out

Section 5.3.4 is valid with a specific feature: In case of bonded fasteners two verifications are to be performed. The first verification is based on combined pull-out and concrete break-out and the second is based on concrete cone failure. Therefore, the verification according to Equation 5.6 is to be replaced with the verification after Equation 6.3.

$$V_{Rk,cp} = min\{k_3 \cdot N_{Rk,p}; \quad k_3 \cdot N_{Rk,c}\} \tag{6.3}$$

with:

$N_{Rk,p}$ see Section 6.2.3
$N_{Rk,c}$ see Section 6.2.4

The peculiarities mentioned in Section 5.3.4 shall be considered.

6.3.4 Concrete edge failure

Section 5.3.5 applies without modification.

6.4 Combined tension and shear

Section 5.4 applies without modification.

7 Verification of ultimate limit state by elastic analysis for headed fasteners

7.1 General

The design procedure of CEN/TS, Part 2 described in the following sections applies to headed fasteners (Figure 2.2) and corresponds in essential points to CEN/TS, Part 4, design method *A* for post-installed mechanical systems (see Section 5).

The design procedures presume that the welded seams for welded connections were produced according to an ETA and the design of welding is in accordance with EN 1993-1 (Eurocode 3). Furthermore it should be observed that the headed fastener is fixed to the formwork or auxiliary constructions in a way that no movement of the fastener will occur during placing of reinforcement or during pouring and compacting of the concrete. The concrete must be adequately compacted particularly under the head of the stud or fastener and under the fixture. Hence, vent openings in fixtures larger than 400 mm × 400 mm are to be present. A fixture not exceeding 200 mm × 200 mm and a number of 4 fasteners can be placed simultaneously during vibrating the concrete. The above remarks concerning the protection of the position of the fastening and the compaction of the concrete apply. The fastenings should not be moved after vibrating has been finished. It is not permitted to place the anchor plates by pushing into the concrete because voids in the concrete adjacent to the headed fasteners cannot be foreclosed.

Inspection and approval of the correct installation of the fasteners shall be carried out by appropriately qualified personnel as in case of post-installed chemical and mechanical systems.

7.2 Tension forces in the supplementary reinforcement

7.2.1 Detailing of supplementary reinforcement in case of tension loaded fastenings

The tension load $N_{Ed,re}$ in the supplementary reinforcement is calculated using Section 4.4. It should be designed to resist the total load acting on the fastening. The supplementary reinforcement should comply with the following requirements:

a) The reinforcement should consist of ribbed reinforcing bars ($f_{yk} \leq 500\ \text{N/mm}^2$) with a diameter d_s not larger than 16 mm and should be detailed in form of stirrups or loops with a mandrel diameter according to EN 1992-1-1. In general, to avoid mixing up different sizes of reinforcement the same diameter of the reinforcement should be provided for all fasteners of a group.
b) The supplementary reinforcement should be placed as close to the fasteners as practicable to minimize the effect of eccentricity associated with the angle of the failure cone. Preferably, the supplementary reinforcement should enclose the surface reinforcement. Only these reinforcement bars with a distance $\leq 0.75\,h_{ef,}$ from the fastener should be assumed as effective.

Design of Fastenings for Use in Concrete – the CEN/TS 1992-4 Provisions. First edition.
Rainer Mallée, Werner Fuchs, Rolf Eligehausen.

c) The minimum anchorage length of supplementary reinforcement in the concrete failure cone is min $l_1 = 4d_s$ (anchorage with bends, hooks or loops) or min $l_1 = 10d_s$ (anchorage with straight bars with or without welded transverse bars).
d) The supplementary reinforcement should be anchored outside the assumed failure cone with an anchorage length l_{bd} according to EN 1992-1-1:2004 (Eurocode 2).
e) A surface reinforcement should be provided as shown in Figure 4.17 designed to resist the splitting forces according to CEN/TS, Part 2, Section 6.2.6.

7.2.2 Detailing of supplementary reinforcement in case of shear loaded fastenings

When the shear force acting on the fasting shall be resisted by a surface reinforcement for example according to Figure 4.18, the requirements for supplementary reinforcement of Section 7.2.1 apply. Nevertheless, the permissible distance of the supplementary reinforcement to the headed bolt does not depend on the embedment depth but on the edge distance. Only bars with a distance $\leq 0.75c_1$ from the fastener should be assumed as effective. If the supplementary reinforcement encloses and contacts the shaft of the fastener and be positioned as closely as possible to the fixture (Figure 4.19), the supplementary reinforcement should be anchored outside the assumed failure cone with an anchorage length $l_{b,d}$ according to EN 1992-1-1:2004 (Eurocode 2).

7.3 Tension load

7.3.1 Required verifications

7.3.1.1 Fastening without supplementary reinforcement

Headed fasteners shall be verified for the same failure modes as post-installed mechanical fasteners (Section 5.2.1). However, headed fasteners can be installed with substantially lower edge distances than post-installed mechanical fasteners. Hence, failure in the area of the head by a local (conical) concrete break-out in the direction of the free edge can occur (Figure 7.1). This failure mode is called local

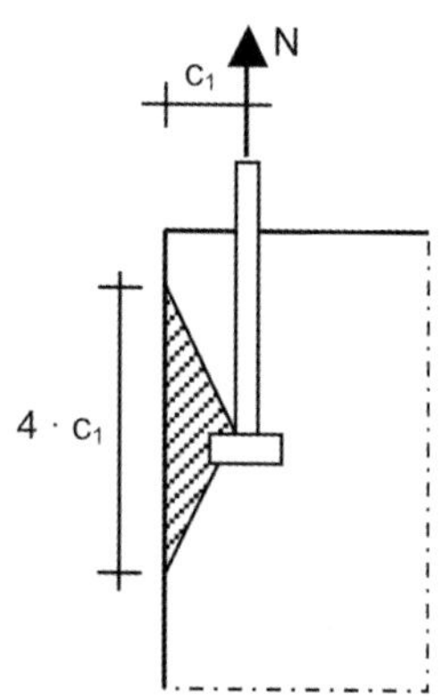

Fig. 7.1 Local concrete blow-out of a headed fastener under tension load with very small edge distance c_1

concrete break-out or blow out. The middle angle of slope towards the vertical is between 20° and 30°, on average about 25°. The depth of the break-out body is assumed the edge distance c_1 for reasons of simplification. Thus the diameter of a break-out body is about 4-times the edge distance c_1. Spacing and edge distances have an important effect on the load-carrying capacity of fastenings with headed bolts in case of blow-out failure. Hence, the verification of fastener groups for this failure mode is to be performed with the complete fastening.

The failure mode local concrete break-out or blow-out is not observed with post-installed mechanical fasteners, because these split the concrete already during the installation if the edge distances are very small. This is prevented by the definition of corresponding minimum edge distances c_{min} for mechanical anchors in the relevant ETA.

7.3.1.2 Fastenings with supplementary reinforcement

The supplementary reinforcement should be detailed and designed to resist the total tension load. Then concrete cone failure needs not to be verified. This verification is replaced by the proof for steel and bond failure of the supplementary reinforcement.

7.3.2 Steel failure

The characteristic resistance of a headed fastener in case of steel failure $N_{Rk,s}$ is given in the relevant ETA. $N_{Rk,s}$ is based on f_{uk} as with post-installed fasteners.

7.3.3 Pull-out failure

The characteristic resistance of a headed fastener $N_{Rk,p}$, is given in the relevant ETA. It is limited by the concrete pressure below the head of the fastener. For a preliminary design $N_{Rk,p}$ can be determined as follows:

$$N_{Rk,p} = 6 \cdot A_h \cdot f_{ck,cube} \cdot \psi_{ucr,N} \tag{7.1}$$

with:

A_h load bearing area of the head of the fastener

$$= \frac{\pi}{4} \cdot \left(d_h^2 - d^2\right) \tag{7.2}$$

d_h diameter of the head [mm]
d diameter of the shaft [mm]
$f_{ck,cube}$ characteristic cube strength of the concrete strength class but noting the limitations given in the relevant ETA
$\psi_{ucr,N}$ = 1.0 for fasteners in cracked concrete
= 1.4 for fasteners in uncracked concrete

7.3.4 Concrete cone failure

In the verification of conical concrete break-out it is assumed that the break-out cone originates at the bottom of the head of the fastener and proceeds in an angle of 35°

measured towards of the horizontals up to the concrete surface (Figure 5.2). Therefore, the same conditions are given as for mechanical fasteners. Thus, in principle Section 5.2.4 applies without modifications. Tests indicated that the ultimate concrete break-out resistance of headed fasteners is higher than for normal post-installed mechanical fasteners. This is attributed to the favourable effect of the head. Therefore according to current experience the value $k_{cr} = 8.5$ in Equation 5.2a$_1$ and $k_{ucr} = 11.9$ in Equation 5.2a$_2$. Otherwise Section 5.2.4 applies without modification.

7.3.5 Splitting

The minimum values for edge distances $c_{\min}$, spacing $s_{\min}$ and member thickness h_{min} of ETA have to be observed to avoid splitting during installation even if the headed fasteners are not torqued to ensure proper casting and compacting of the concrete.

Verification for splitting of loaded headed fasteners is performed in accordance with Section 5.2.5.2. In addition the following Equation 7.3 to determine the required cross-section A_s of the splitting reinforcement is given:

$$A_s = 0.5 \cdot \frac{\sum N_{Ed}}{f_{yk}/\gamma_{Ms,re}} \quad [\mathrm{mm}^2] \tag{7.3}$$

with:

$\sum N_{Ed}$ sum of the design tensile force of the fasteners in tension under the design value of the actions [N]

f_{yk} nominal yield strength of the reinforcing steel ≤ 500 N/mm^2

In principle this approach may be also used for post-installed fasteners. However, depending on the type of fastener the splitting force of post-installed mechanical fastener with the identical tension load $\sum N_{Ed}$ amounts to twice to three times the value of a headed fastener.

7.3.6 Local concrete break-out (blow-out)

Verification of blow-out failure is not required if the edge distance in all directions exceeds $c = 0.5\,h_{\mathrm{ef}}$. Then blow-out failure does not govern.

For groups of fasteners perpendicular to the edge, which are loaded uniformly, verification is only required for the fasteners closest to the edge.

The characteristic resistance in case of blow-out failure is:

$$N_{Rk,cb} = N^0_{Rk,cb} \cdot \frac{A_{c,Nb}}{A^0_{c,Nb}} \cdot \psi_{s,Nb} \cdot \psi_{g,Nb} \cdot \psi_{ec,Nb} \cdot \psi_{ucr,N} \quad [\mathrm{N}] \tag{7.4}$$

The individual factors of Equation 7.4 are explained in the following

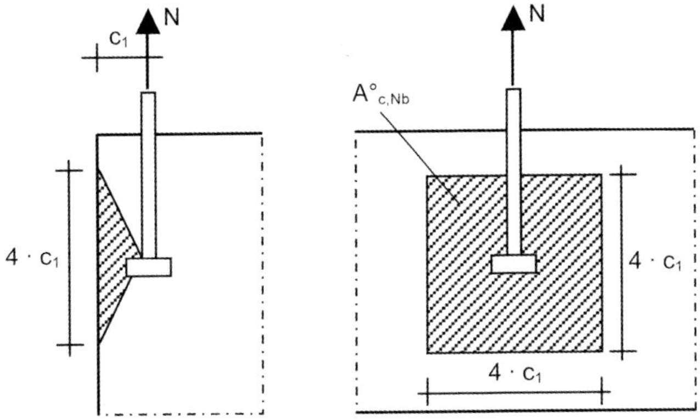

Fig. 7.2 Idealization of the concrete break-out body of a single headed fastener with local concrete blow-out and projected area $A^0_{c,Nb}$

7.3.6.1 Characteristic resistance of a single headed fastener

The characteristic resistance of a single headed fastener not influenced by adjacent fasteners or free structural component edges placed in cracked concrete after Furche and Eligehausen (1991) is obtained by:

$$N^0_{Rk,cb} = 8 \cdot c_1 \cdot \sqrt{A_h} \cdot \sqrt{f_{ck,cube}} \quad [\mathrm{N}] \tag{7.4a}$$

with:

$f_{ck,cube}$ characteristic cube strength of the concrete strength class but noting the limitations given in the relevant ETA
A_h load bearing area of the head of the fastener, see Equation 7.2 [mm^2]
c_1 edge distance, see Figure 7.2 [mm]

7.3.6.2 Effect of spacing and further edge distances

The geometric effect of axial spacing and further edge distances (e.g. fastening in a corner (Figure 7.3b) or in a thin concrete member (Figure 7.3c)) on the characteristic resistance is taken into account following the same systematics as in case of concrete cone failure by the value $A_{c,Nb}/A^0_{c,Nb}$. Governing parameter is here, however, the edge distance c_1:

$A^0_{c,Nb}$ reference projected area, see Figure 7.2

$$= (4 \cdot c_1)^2 \tag{7.4b}$$

$A_{c,Nb}$ actual projected area, limited by overlapping concrete break-out bodies of adjacent fasteners ($s \leq 4\ c_1$) as well as by edges of the concrete member ($c_2 \leq 2 \cdot c_1$) or the member depth

Examples for the calculation of $A_{c,Nb}$ are given in Figure 7.3.

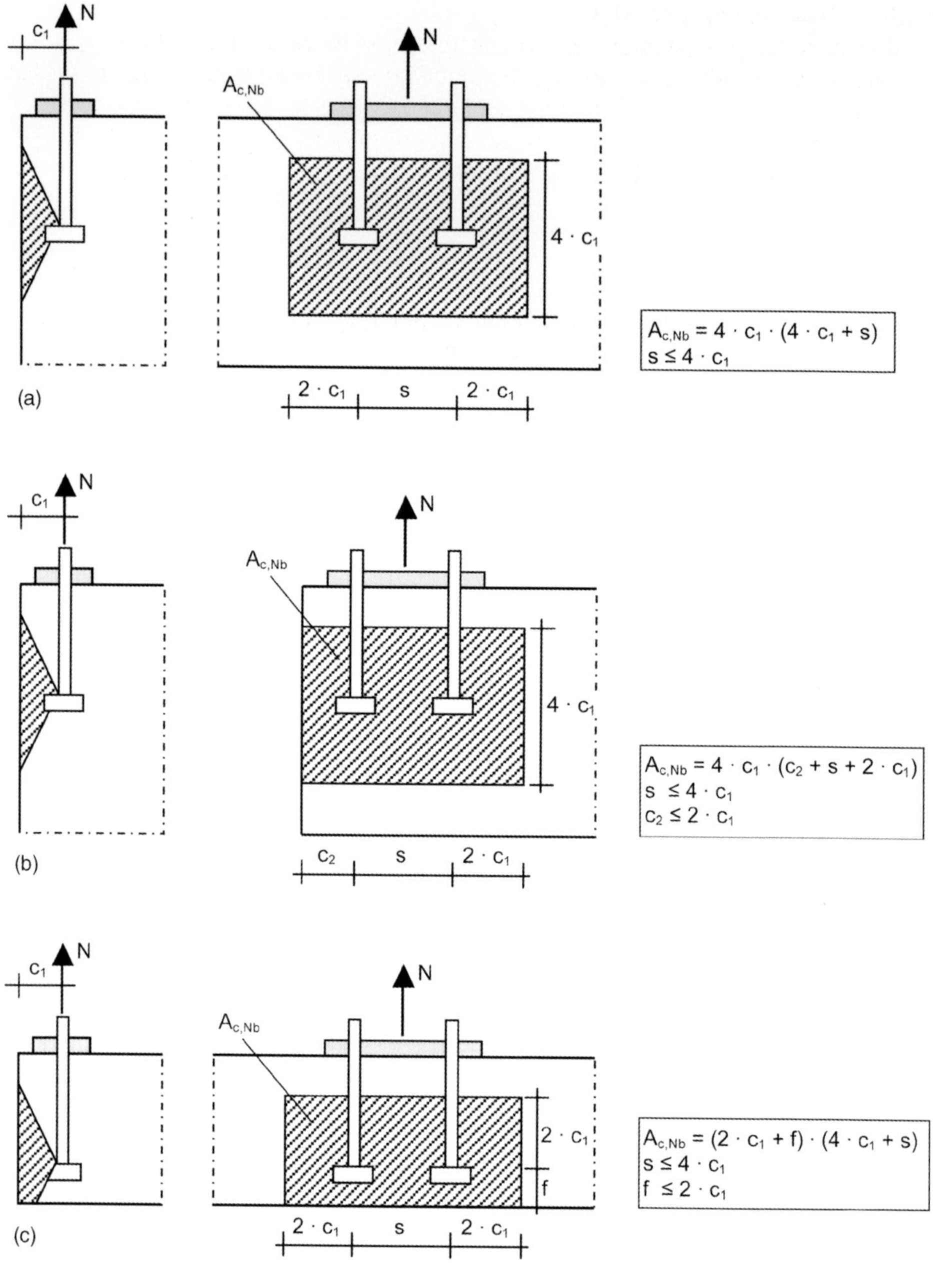

Fig. 7.3 Examples of areas $A_{c,Nb}$ for idealized concrete break-out bodies with the failure mode local concrete blow-out

(a) double fastening with headed fasteners in a thick component

(b) double fastening with headed fasteners in the corner of a thick component

(c) double fastening with headed fasteners in a thin component

7.3.6.3 Free component edges

Analogous to the procedure in case of concrete cone break-out the factor $\psi_{s,Nb}$ takes account of the disturbance of the distribution of stresses in the concrete due to further edges of the concrete member (e.g. fastening in a corner of the concrete member, see Figure 7.3b). For fastenings with several edge distances (e.g. fastening in a narrow concrete member), the smallest edge distance, c_2, should be inserted in Equation 7.4.

$$\psi_{s,Nb} = 0.7 + 0.3 \cdot \frac{c_2}{c_1} \leq 1 \tag{7.4c}$$

It should be noted that the text is ambiguous in CEN/TS because the factor $\psi_{s,Nb}$ is influenced only by edges vertically to the direction of c_1.

7.3.6.4 Effect of the bearing area on the behaviour of groups

The factor $\psi_{g,Nb}$ takes account of the effect small spacing of the load bearing areas on the individual headed fasteners in a group on its characteristic resistance.

$$\psi_{g,Nb} = \sqrt{n} + (1 - \sqrt{n}) \cdot \frac{s_1}{4 \cdot c_1} \geq 1 \tag{7.4d}$$

with:

n number of tensioned fasteners in a row parallel to the edge
s_1 spacing $\leq 4 \cdot c_1$

The factor decreases with $\psi_{g,Nb} = \sqrt{n}$ for $s = 0$ to $\psi_{g,Nb} = 1$ for $s = 4 \cdot c_1$. This can be explained as follows:

If the spacing of the tensioned headed fasteners in a row with n fasteners is reduced to $s = d_h \approx 0$, the load bearing area of the fictive headed fastener is $n \cdot A_h$. The resistance is proportional to the load bearing area (Equation 7.4a). Hence, the resistance of the fictive headed fastener amounts to $\sqrt{n}$-times of a headed fastener bolt with the load bearing area A_h. For a spacing $s = 4 \cdot c_1$ the resistance of the group is n-times the resistance of a single headed fastener. Between these limits it is interpolated linearly.

7.3.6.5 Effect of load eccentricity

The factor $\psi_{ec,Nb}$ takes account of a group effect, when different loads are acting on the individual fasteners of a group for example due to loading by means of a moment.

$$\psi_{ec,Nb} = \frac{1}{1 + 2 \cdot e_N/(4 \cdot c_1)} \leq 1 \tag{7.4e}$$

with:

e_N eccentricity of the resulting tensile load in respect of the centre of gravity of the tensioned fasteners (see Figure 5.12)

7.3.6.6 Effect of the position of the fastening

The effect of the position of the fastening in cracked or non-cracked concrete is considered by the factor $\psi_{ucr,N}$:

$$\psi_{ucr,N} = 1.0 \text{ for headed fasteners in cracked concrete} \quad (7.4f_1)$$

$$\psi_{ucr,N} = 1.4 \text{ for headed fasteners in cracked concrete} \quad (7.4f_2)$$

7.3.7 Steel failure of the supplementary reinforcement

The characteristic resistance of reinforcement $N_{Rk,re}$ supplementary to a fastening with headed fasteners is:

$$N_{Rk,re} = n \cdot A_s \cdot f_{yk} \quad (7.5)$$

with:

n number of legs of the supplementary reinforcement effective for one fastener
A_s cross section of one leg of the supplementary reinforcement
f_{yk} nominal yield strength of the supplementary reinforcement $\leq 500\ \text{N/mm}^2$

7.3.8 Anchorage failure of the supplementary reinforcement in the concrete cone

The design resistance $N_{Rd,a}$ of the supplementary reinforcement of one headed fastener is given by:

$$N_{Rd,a} = \sum_n \frac{l_1 \cdot \pi \cdot d_s \cdot f_{bd}}{\alpha} \quad (7.6)$$

with:

l_1 anchorage length of the supplementary reinforcement in the assumed failure cone (see Figure 4.17) $\geq l_{b,min}$
$l_{b,min}$ minimum anchorage length of the rebar
$= 4 \cdot d_s$ (anchorage with bends, hooks or loops)
$= 10 \cdot d_s$ (anchorage with straight bars with or without welded transverse bars)
d_s diameter of the reinforcement bar
f_{bd} design bond strength according to EN 1992-1-1:2004 (Eurocode 2) taking into account the concrete cover of the supplementary reinforcement
α influencing factor, according to EN 1992-1-1:2004 (Eurocode 2)
$= 0.7$ for hooked bars
n number of legs of the supplementary reinforcement effective for one fastener

The determination of the design resistance of the supplementary reinforcement in case of bond failure in the break-out cone is conforming to the provisions in EN 1992-1-1:2004 (Eurocode 2) for rebars under tension loading.

7.4 Shear load

7.4.1 Required verifications

7.4.1.1 Fastenings without supplementary reinforcement

For headed fasteners under shear load without supplementary reinforcement the same failure modes are observed as for post-installed mechanical fasteners. Therefore, the verifications of steel failure are to be performed for shear loads without and with lever arm, of concrete pry-out and of concrete edge failure.

7.4.1.2 Fastenings with supplementary reinforcement

The verifications for shear loaded headed fasteners with supplementary reinforcement the proofs of steel failure are to be carried out for shear forces without and with lever arm, for concrete pry-out as well as for steel and anchorage failure of the supplementary reinforcement. Concrete edge failure is not to be proved because it is assumed that the supplementary reinforcement takes up the entire shear load acting on the fastening.

7.4.2 Steel failure of the headed fastener

Sections 5.3.2 and 5.3.3 apply.

7.4.3 Concrete pry-out failure

Section 5.3.4 applies taking into account the values for k_{cr} and k_{ucr} after Section 7.3.4.

7.4.4 Concrete edge failure

Section 5.3.5 is valid without modification. In case of attachments with headed fasteners in a narrow thin concrete component it is favourable to assign the forces always to a supplementary reinforcement.

7.4.5 Steel failure of the supplementary reinforcement

The characteristic resistance of one fastener in case of steel failure of the supplementary reinforcement may be calculated according to Equation 7.7.

$$N_{Rk,re} = k_6 \cdot n \cdot A_s \cdot f_{yk} \tag{7.7}$$

with:

k_6 efficiency factor
  = 1.0 surface reinforcement according to Figure 4.18
  = 0.5 supplementary reinforcement according to Figure 4.19
n number of bars of the supplementary reinforcement of one fastener
A_s cross section of one bar of the supplementary reinforcement
f_{yk} nominal yield strength of the supplementary reinforcement $\leq 500\,N/mm^2$

The factor $k_6 = 0.5$ for supplementary reinforcement according to Figure 4.19 takes account of unavoidable tolerances in workmanship on site.

7.4.6 Anchorage failure of the supplementary reinforcement in the concrete break-out body

The verification for anchorage failure in the concrete break-out body does not have to be carried out for supplementary reinforcement detailed according to Figure 4.19. Nevertheless, the anchorage in the concrete member is to be verified. If a supplementary reinforcement is detailed as surface reinforcement (Figure 4.18), the design resistance $N_{Rd,a}$ of the supplementary reinforcement is determined according to Section 7.3.8.

7.5 Combined tension and shear load

Section 5.4 applies without modification for fastenings with headed fasteners without supplementary reinforcement and fastenings with supplementary reinforcement to take up the tension and shear forces.

For fastenings with supplementary reinforcement which is able to resist only tension *or* shear loads the following interaction is to be used:

$$\beta_N^{k_7} + \beta_V^{k_7} \leq 1 \tag{7.8}$$

In Equation 7.8 the largest value of β_N and β_V for the different failure modes should be inserted. The value k_7 is given in the relevant ETA. According to current experience $k_7 = 2/3$.

fib (2011) recommends in case of fastenings with headed fasteners to carry out separate interaction proofs of steel failure of the headed fasteners and concrete failure (Section 5.4). Then the verification of concrete failure for fastenings with supplementary reinforcement is to be performed after Equation 7.8 and steel and anchorage failure of the supplementary reinforcement are to be taken into account in the calculation of β_N and β_V. This approach should be considered in the revision of CEN/TS.

8 Verification of ultimate limit state by elastic analysis for anchor channels

8.1 General

The design procedure of CEN/TS, Part 3 described in the following sections, is valid for anchor channels according to Figure 2.3. It is based on the same mechanical models and the same systematics as the CC-method for post-installed mechanical and headed fasteners, nevertheless, was modified for anchor channels. In cases of the failure modes concrete cone break-out and local concrete break-out (blow out) under tensile loading as well as concrete edge failure and concrete pry-out under shear load it is not calculated the load-carrying capacity of the fastener group - like with post-installed mechanical and headed fasteners – but the capacity of *one* anchor. This is justified as follows:

In case of fastenings with post-installed and headed fasteners a stiff base plate is assumed. Therefore the forces to be resisted by the individual fasteners of a group of fasteners are determined in accordance with the theory of elasticity (see Section 4.2.1). This approach yields a linear distribution of the forces acting eccentrically on the fixture (Figure 4.2). The distribution of fastener forces is taken into account by the value $\psi_{ec,N}$ in the calculation of the characteristic concrete break-out resistance of a group with post-installed and headed fasteners.

On the other hand anchor channels behave like continuous beams supported by the anchors elastically and having a partial end restraint. Furthermore differently high tension and shear loads can act at arbitrary positions of the anchor channel. The distribution of loads to be resisted by the single anchors of the channel can be very irregular (Figure 8.1). Therefore the CC-method for post-installed and headed fasteners does not apply to anchor channels with more than two anchors. The calculation of the forces to be resisted by the anchors of an anchor channel is given in Section 4.2.2 for tension loading and in Section 4.3.2 for shear loading.

The following assumptions in respect to installation of anchor channels shall be satisfied. The installation instructions of the manufacturers should reflect them in detail:

- anchor channels should be fixed to the formwork or auxiliary constructions in a way that no movement of the anchor channel will occur during placing of reinforcement or during pouring and compacting of the concrete.
- the concrete should be adequately compacted particularly under the head of the anchor and under the channel.
- it is accepted to vibrate the anchor channels into the wet concrete immediately after pouring. Anchor channels with a length of smaller than 1m are allowed to be placed by one installer, longer anchor channels should be placed by at least two persons to ensure simultaneous placing into the concrete. To avoid cavities in the concrete underneath the channel the concrete in the region of the anchor and the anchor

Design of Fastenings for Use in Concrete – the CEN/TS 1992-4 Provisions. First edition.
Rainer Mallée, Werner Fuchs, Rolf Eligehausen.

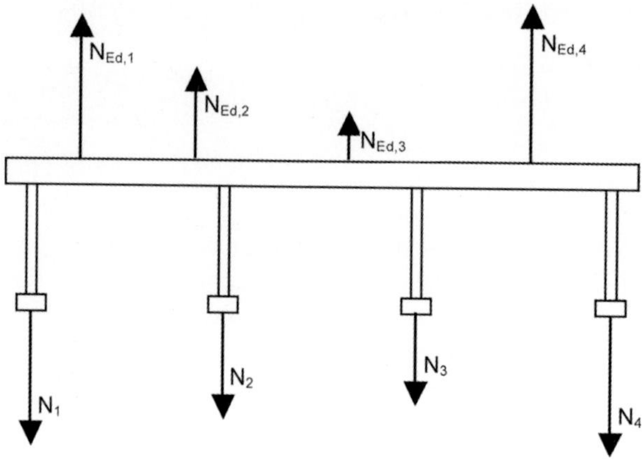

Fig. 8.1 Distribution of the loads acting on the anchors of an anchor channel under tension load

channel shall be properly compacted. In addition, the anchor channels shall not be moved after vibrating has been finished.

- the correct installation of anchor channels shall be performed by appropriately qualified personnel. This applies especially to vibrated anchor channels. This is also valid for inspection and approval of the correct installation.

As a basic principle placing anchor channels by only pushing them into the wet concrete is not allowed because voids in the concrete underneath the channel and adjacent to the heads of the anchors are to be expected. This would cause a significant reduction of the load bearing capacity of the anchor channel.

In the design of anchor channels it is distinguished - as with post-installed mechanical fasteners (design method *A*) and headed fasteners - in load directions and failure modes. If anchor channels are subjected to tension forces, in addition to the failure modes shown in Figure 5.1 the following failures can occur (Figure 8.2):

- local failure of the channel lip
- failure of the anchor or of the connection between anchor and channel
- failure due to flexure of the channel.

The above mentioned failure modes can also occur in case of shear loading with the exception of failure due to flexure of the channel. Then, this failure mode is prevented by the adjoining concrete.

The following actions are not covered by CEN/TS:

- shear in the direction of the longitudinal axis of the channel
- fatigue loading
- seismic loading.

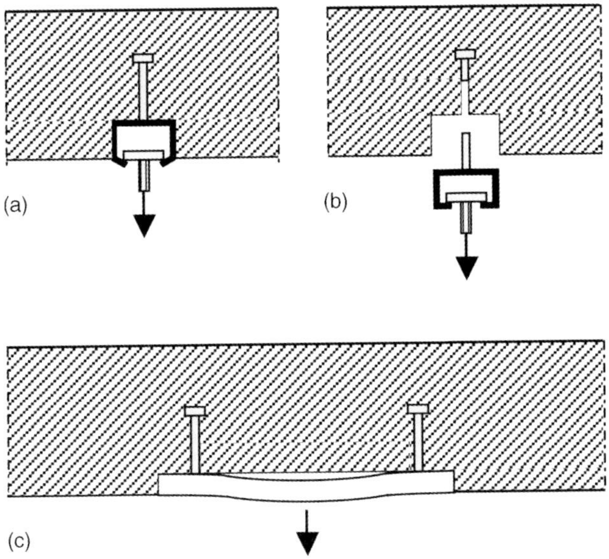

Fig. 8.2 Additional failure modes of tension loaded anchor channels (rupture by failure of the connection between anchor and channel is not shown) (Wohlfahrt (1996))
(a) local flexure of the anchor channel lip
(b) failure of the anchor
(c) failure due to flexure of the anchor channel

The design method described in the following applies exclusively to anchor channels with a ETA. It is based in the essentials on the results of the investigations of Wohlfahrt (1996), Kraus (2003) and Potthoff (2008). The results of these investigations are summarized in Eligehausen *et al.* (Eligehausen *et al.* (2007)) where the load bearing behaviour and the design of anchor channels is explained in detail.

8.2 Tension forces in the supplementary reinforcement

8.2.1 Detailing of supplementary reinforcement in case of tension loaded anchor channels

If the failure mode concrete break-out governs, the load-carrying capacity of the anchor channel can be increased by supplementary reinforcement. The tension force $N_{Ed,re}$ in the reinforcement is calculated for the highest loaded anchor (see Section 4.4) for the tension force N_{Ed}^{a} acting on the anchor. The same supplementary reinforcement should be used for all anchors to avoid confusion on the building site. The supplementary reinforcement must be detailed and anchored as for fastenings with headed fasteners (see Section 7.2.1 and Figure 8.3). If anchor channels are arranged in parallel with the edge of a concrete member or in a narrow component, the levels of the reinforcement must be oriented perpendicular to the longitudinal axis because during the loading a crack forms beneath the anchor channel in the longitudinal direction of the anchor

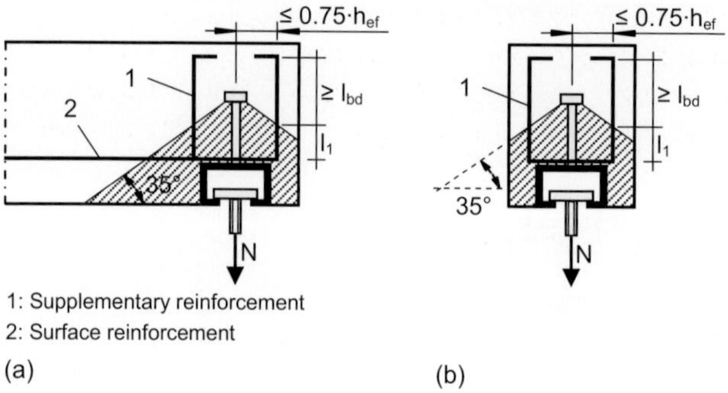

Fig. 8.3 Detailing of supplementary reinforcement to resist tension load
(a) in the component edge
(b) in a narrow component

channel and, hence, a reinforcement positioned in parallel with the anchor channel axis is not effective (Kraus (2003)).

8.2.2 Detailing of supplementary reinforcement in case of shear loaded anchor channels

The load-carrying capacity of anchor channels can be increased by a supplementary reinforcement according to Figure 4.18 in the form of stirrups and straight edge reinforcement in case of the failure mode concrete edge failure. On the other hand a supplementary reinforcement according to Figure 4.19 is little effective in case of anchor channels for reasons already mentioned in Section 4.4 and, hence, not covered in CEN/TS. The text in Section 8.2.1 with respect to detailing and placing of the supplementary applies accordingly. The tension force N_{Ed}^{a} in the reinforcement according to Section 4.4 is calculated for the highest loaded anchor. Anchor channels transfer a significant share of the shear force directly into the concrete via the channel (see Section 4.3.2). Hence, the supplementary reinforcement should be designed for the highest shear force acting on the anchor channel if this value is bigger than the highest shear load calculated for the anchor (Schmid (2010)). Furthermore at least one rebar of the supplementary reinforcement should be placed in the area of the acting shear force. The distance to the shear load should be $\leq 0.75 \cdot c_1$.

8.3 Tension load

8.3.1 Required verifications

8.3.1.1 Anchor channels without supplementary reinforcement

For anchor channels under tension load without supplementary reinforcement the same verifications shall be performed as for headed fasteners (see Section 7.3.1). However, since anchor channels can experience further failure modes (Figure 8.2), the failure modes rupture of the special screw (channel bolt), local failure of the channel lip, failure of the connection between anchor and channel as well as failure of the anchor shall be verified.

8.3.1.2 Anchor channels with supplementary reinforcement

If a supplementary reinforcement is detailed and designed according to Section 7.2.1 and Figure 8.3 concrete cone failure needs not to be verified. This verification is replaced by the proof for steel and bond failure of the supplementary reinforcement as for headed fasteners.

8.3.2 Steel failure of channel bolt and channel

The characteristic values of the resistance $N_{Rk,s,a}$ (rupture of the anchor), $N_{Rk,s,c}$ (failure of the connection between anchor and channel), $N_{Rk,s,l}$ (local failure of the channel lip), $N_{Rk,s}$ (rupture of the special screw (channel bolt)) and $M_{Rk,s,flex}$ (failure due to flexure of the channel) are given in the relevant ETA. The values $N_{Rk,s,a}$ and $N_{Rk,s}$ are calculated according to Equation 5.1. The remaining values are determined from the results of product prequalification tests. In general ETA states only the minimum values $N_{Rk,s,a}$ and $N_{Rk,s,c}$.

8.3.3 Pull-out failure

Section 7.3.3 applies.

8.3.4 Concrete cone failure

Concrete cone break-out can occur in case of short anchors as well as of small spacing and/or edge distances and limits the load-carrying capacity of the anchor channel. In principle, with this failure mode the load-bearing behaviour corresponds to the performance of post-installed and headed fasteners. Indeed, the anchor channel embedded in the concrete break-out body can unfavourably affect the resistance according to the relation of profile height to embedment depth (Kraus (2003)). The channel disturbs the distribution of the stresses in the concrete and causes tensile stresses in the concrete by the obstruction of the displacements due to shrinkage of the adjacent surface concrete by the anchors. This effect is considered by the reduction factor α_{ch}. Furthermore, for reasons mentioned in Section 8.1 the characteristic resistance is not calculated for the fastener group but for one anchor. The anchor with the highest ratio $N_{Ed}^{a}/N_{Rk,c}$ governs the design. Loading as well as resistance of the individual anchors of an anchor channel can be different. Therefore, all anchors are to be verified if necessary.

The characteristic resistance of one anchor of an anchor channel is (Kraus (2003)):

$$N_{Rk,c} = N_{Rk,c}^{0} \cdot \alpha_{s,N} \cdot \alpha_{e,N} \cdot \alpha_{c,N} \cdot \psi_{re,N} \cdot \psi_{ucr,N} \tag{8.1}$$

The different parameters in Equation 8.1 are explained in the following:

8.3.4.1 Characteristic resistance of a single anchor

The basic characteristic resistance of one anchor not influenced by adjacent anchors, edges or corners of the concrete member located in cracked concrete is obtained by:

$$N_{Rk,c}^{0} = 8.5 \cdot \alpha_{ch} \cdot \sqrt{f_{ck,cube}} \cdot h_{ef}^{1,5} \quad [\text{N}] \tag{8.1a}$$

with:

α_{ch} factor taking into account the influence of the channel on the concrete cone failure load ≤ 1

$f_{ck,cube}$ characteristic cube strength of the concrete strength class but noting the limitations given in the relevant ETA

h_{ef} embedment depth [mm]

The factor α_{ch} is given in the relevant ETA. It is calculated for anchor channels with $h_{ch}/h_{ef} \leq 0.4$ and $b_{ch}/h_{ef} \leq 0.7$ according to Equation 8.1b.

$$\alpha_{ch} = (h_{ef}/180)^{0.15} \leq 1 \tag{8.1b}$$

Equation 8.1b is based on numeric analysis confirmed by test results (Kraus (2003)).

8.3.4.2 Effect of neighbouring anchors

The influence of neighbouring anchors on the concrete cone resistance is taken into account by the factor $\alpha_{s,N}$ according to Equation 8.1c:

$$\alpha_{s,N} = \frac{1}{1 + \sum_{i=1}^{n} \left[\left(1 - \frac{s_i}{s_{cr,N}}\right)^{1,5} \cdot \frac{N_i}{N_0} \right]} \tag{8.1c}$$

with (see Figure 8.4):

s_i distance between the anchor under consideration and the neighbouring influencing anchor i
$\leq s_{cr,N}$

$$s_{cr,N} = 2 \cdot (2.8 - 1.3 \cdot h_{ef}/180) \cdot h_{ef} \geq 3 \cdot h_{ef} \tag{8.1d}$$

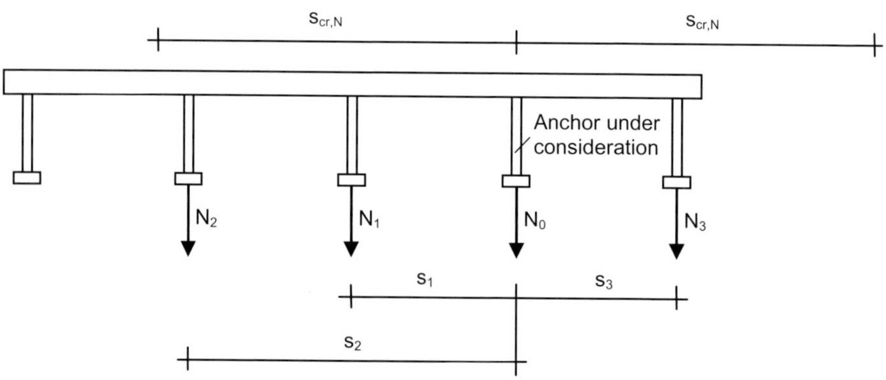

Fig. 8.4 Example of an anchor channel loaded by different tension forces (Kraus (2003))

N_i tension force of an influencing anchor i
N_0 tension force of the anchor under consideration
n number of anchors within a distance $s_{cr,N}$ to both sides of the anchor under consideration

The factor $\alpha_{s,N}$ of Equation 8.1c replaces the ratio $A_{c,N}/A^0_{c,N}$ and the factor $\psi_{ec,N}$ in Equation 5.2 for the calculation of the characteristic resistance of fastenings with post-installed and headed fasteners in case of concrete break-out failure. $\alpha_{s,N}$ is explained by means of Figure 8.5 showing an anchor channel with two anchors. The concrete break-out resistance of anchor 1 in Figure 8.5a is influenced by the distance to the

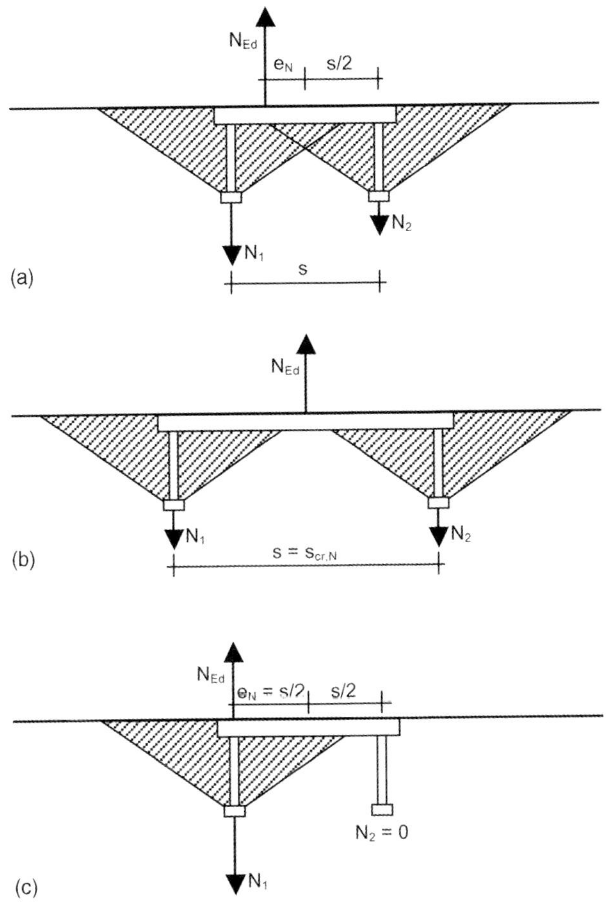

Fig. 8.5 Influence of the spacing and the anchor loading on the concrete break-out body (Eligehausen *et al.* (2007))
(a) spacing $s < s_{cr,N}$, both anchors effective
(b) spacing $s = s_{cr,N}$, both anchors effective
(c) spacing $s < s_{cr,N}$, only left anchor effective

anchor 2 and its loading. If the distance of the anchor is so large that the break-out cones do not overlap ($s \geq s_{cr,N}$), anchor 2 does not influence the concrete break-out resistance of anchor 1 (Figure 8.5b). This is also valid for $s < s_{cr,N}$ if anchor 2 is not loaded (Figure 8.5c). The influence of the spacing and of the load acting on the neighbouring anchor is taken into account by multiplication of the characteristic resistance $N^0_{Rk,c}$ with the value $\alpha_{s,N}$. For the example in Figure 8.4 it is:

$$\alpha_{s,N} = \frac{1}{1 + \left(1 - \frac{s_1}{s_{cr,N}}\right)^{1.5} \cdot \frac{N_1}{N_0} + \left(1 - \frac{s_2}{s_{cr,N}}\right)^{1.5} \cdot \frac{N_2}{N_0} + \left(1 - \frac{s_3}{s_{cr,N}}\right)^{1.5} \cdot \frac{N_3}{N_0}}$$

For anchor channels with two anchors the calculations with the factor $\alpha_{s,N}$ and the product $(A_{c,N}/A^0_{c,N}) \cdot \psi_{ec,N}$ yield practically the same result if for both fastenings the same value of $s_{cr,N}$ is presumed.

While it is supposed with post-installed and headed fasteners that the characteristic spacing always corresponds to the value $3 \cdot h_{ef}$ (see Section 5.2.4), investigations by Kraus (2003) have indicated that with anchor channels it is explicitly dependent on the embedment depth. The characteristic spacing varies between $s_{cr,N} = 5 \cdot h_{ef}$ (for $h_{ef} = 40$ mm) and $s_{cr,N} = 3 \cdot h_{ef}$ (for $h_{ef} \geq 180$ mm).

8.3.4.3 Effect of edges of the concrete member

The influence of an edge of the concrete member on the characteristic resistance is taken into account by the factor $\alpha_{e,N}$ according to Equation 8.1e:

$$\alpha_{e,N} = \left(\frac{c_1}{c_{cr,N}}\right)^{0,5} \leq 1 \qquad (8.1e)$$

with:

c_1 edge distance of the anchor channel (see Figure 8.6a)
$c_{cr,N}$ characteristic edge distance

$$\begin{aligned} &= 0.5 \cdot s_{cr,N} \\ &= (2.8 - 1.3 \cdot h_{ef}/180) \cdot h_{ef} \geq 1.5 \cdot h_{ef} \end{aligned} \qquad (8.1f)$$

With anchor channels located in a narrow concrete member with different edge distances $c_{1,1}$ and $c_{1,2}$ (Figure 8.6b) the minimum value of $c_{1,1}$ and $c_{1,2}$ shall be inserted in Equation 8.1e.

If anchor channels are placed in parallel with the component edge (Figure 8.6a) a crack forms under the channel in its longitudinal direction during loading. Hence, the concrete on the side facing away from the edge is activated only over a width according to the edge distance. On the other hand an activation of the concrete over a width according to the characteristic edge distance is assumed in the CC-method for post-installed and headed fasteners.

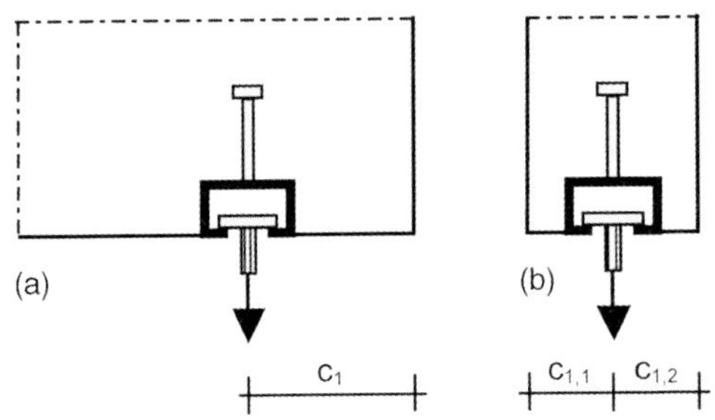

Fig. 8.6 Anchor channel:
(a) in the component edge
(b) in a narrow component

8.3.4.4 Effect of a corner of the concrete member

The influence of a corner of the concrete member on the characteristic resistance is taken into account by the factor $\alpha_{c,N}$ according to Equation (8.1g):

$$\alpha_{c,N} = \left(\frac{c_2}{c_{cr,N}}\right)^{0.5} \leq 1 \tag{8.1g}$$

with:

c_2 corner distance of the anchor under consideration (see Figures 8.7a, 8.7b und 8.7d)

If an anchor is influenced by two corners (see example in Figure 8.7c), then the factor $\alpha_{c,N}$ has to be calculated for the values $c_{2,1}$ and $c_{2,2}$ and the product of the factors $\alpha_{c,N}$ should be inserted in Equation 8.1.

Tests with anchor channels in a component corner do not exist. Nevertheless, it is assumed that the factor $\alpha_{c,N}$ after Equation 8.1g is conservative due to the fact that it was chosen in analogy to the factor $\alpha_{e,N}$.

In the example in Figure 8.7d for the calculation of the factor $\alpha_{c,N}$ the corner distance c_2 and for the factor $\alpha_{e,N}$ the minimum of the values $c_{1,1}$ and $c_{1,2}$ as well as for the determination of the value $\alpha_{c,N}$ the edge distance c_2 is used.

8.3.4.5 Effect of dense surface reinforcement (shell spalling)

The effect of dense surface reinforcement is taken into account by the factor $\psi_{re,N}$. It agrees with the factor used for post-installed and headed fasteners (Equation 5.2).

8.3.4.6 Effect of the anchor channel position

If the anchor channel is anchored in uncracked concrete (verification according to Equation 2.1 is fulfilled), the factor $\psi_{ucr,N} = 1.4$. In principle, it corresponds to the value for post-installed and headed fasteners. However, in case of post-installed and headed

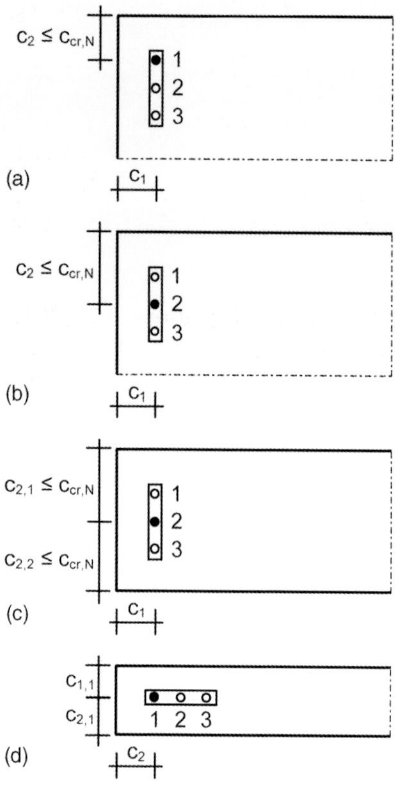

Fig. 8.7 Definition of the edge distance for an anchor channel under tension load located in the component corner
(a) calculation of the resistance of anchor 1
(b) calculation of the resistance of anchor 2
(c) calculation of the resistance of anchor 2
(d) calculation of the resistance of anchor 1

fasteners the effect of the position of the fastening is not taken into account by a factor $\psi_{ucr,N}$ but the different values k_{cr} and k_{ucr} in Equation 5.2.

8.3.4.7 Effect of a narrow member

For the case of anchor channels with $h_{ef} > 180$ mm in an application with influence of neighbouring anchors and influence of an edge and 2 corners (Figure 8.7c) located with edge distance $< c_{cr,N}$ and spacing $< s_{cr,N}$ from the anchor under consideration the calculation according to Equation 8.1 leads to conservative results. More precise results are obtained if the value h_{ef} is substituted by the maximum of Equations 8.1h$_1$ and 8.1h$_2$:

$$h'_{ef} = \frac{c_{\max}}{c_{cr,N}} \cdot h_{ef} \geq 180\,\text{mm} \tag{8.1h$_1$}$$

$$h'_{ef} = \frac{s}{s_{cr,N}} \cdot h_{ef} \geq 180\,\text{mm} \tag{8.1h$_2$}$$

with:

c_{max} maximum distance from the centre of an anchor to the edge of the concrete member (in the example shown in Figure 8.7c c_{max} is the maximum of c_1, $c_{2,1}$ and $c_{2,2}$) $\leq c_{cr,N}$

s maximum centre to centre spacing of anchors measured from the axis of the anchor under consideration $\leq s_{cr,N}$

The value h'_{ef} is inserted in Equation 8.1a as well as in the Equations 8.1b, 8.1d and 8.1f. Reasons for this rule are established in Section 5.2.4.5. Checks demonstrate that in case of the failure mode concrete break-out the characteristic resistance of anchor channels with an embedment depth $h_{ef} \leq 180$ mm is calculated accurately also in the case of application shown Figure 8.7c if h_{ef} is used in the calculation since characteristic spacing and edge distance increase with decreasing embedment depth.

8.3.5 Splitting of the concrete

If the anchor channel is flush with the concrete surface, the concrete is not loaded while tightening the channel bolt (special screw) because the attachment is pulled taut to the channel lips. Nevertheless, in practice the concrete surface is mostly uneven and it cannot be excluded that the anchor channel is located below the concrete surface due to installation inaccuracies. In this case splitting forces develop in the concrete while torqueing the special screw.

Splitting failure is avoided during installation of the special screw by complying with minimum values for edge distance $c_{\min}$, spacing $s_{\min}$, member thickness $h_{\min}$ and requirements on reinforcement as given in the relevant ETA as in case of post-installed fasteners. The minimum values for edge distance depend on channel size and diameter of the channel bolt.

For the verification of splitting failure due to loading Section 5.2.5.2 applies. In general, this verification is not required since the anchor heads are sufficiently large and little splitting forces develop.

8.3.6 Blow-out failure

The equation to calculate the characteristic resistance in case of blow-out given in CEN/TS, Part 3 is based on the model for headed fasteners (see Section 7.3.6), Equation 7.4, however was modified due to the reasons given in Section 8.1. Verification of blow-out failure is not required with anchors when the distance between the anchorage area and the side surface of the structural component exceeds $c > 0.5\,h_{ef}$. Therefore this verification is not further considered.

8.3.7 Steel- and anchorage failure of the supplementary reinforcement

Sections 7.3.7 and 7.3.8 for headed fasteners apply. In any case the characteristic resistance of the supplementary reinforcement is to be determined for one anchor and to be compared with the tension force acting on this anchor.

8.4 Shear loads

8.4.1 Required verifications

Anchor channels without supplementary reinforcement must be verified for the failure modes steel rupture of the channel bolt (special screw), local flexure of channel lip, concrete pry-out, and concrete edge failure. For anchor channels with supplementary reinforcement according to Section 8.2.2 taking the entire external shear force the verification of concrete edge failure is substituted by the verification for steel and anchorage failure of the supplementary reinforcement.

According to CEN/TS, Part 3, in case of a shear force acting with lever arm local failure of the channel does not need to be verified because the characteristic resistance $V_{Rk,s}$ of the special screw significantly decreases with increasing lever arm (Equation 5.5). Indeed, in principle it is not to be excluded that by use of a special screw with the largest usable diameter and a small lever arm the value $V_{Rk,s}$ is bigger than the characteristic resistance $V_{Rk,s,l}$ for failure by local flexure of channel lip. Hence, it is advisable to always perform the verification for failure by local flexure of channel lip.

A verification for failure of the anchors as well as of the connection between anchor and channel is not required in CEN/TS because in the tests carried out up to now a failure of the anchors did not appear. Indeed, this failure mode cannot to be excluded basically because the anchors must take up shear and tension forces whose level depends on the anchor channel geometry (see Section 4.3.2). Moreover, failure of the anchors or of the connection is observed in case of combined tensile and shear whereas the tension resultant of the ultimate resistance is smaller than the resistance under axial tension load. Hence, it is recommended to verify the failure modes rupture of the anchors as well as the connection between anchor and channel under shear loads and combined tension loads and shear loads. The characteristic shear resistance for anchors and the connection between anchor and channel is not given in the ETA up to now. It is conservative to assume the value $V_{Rk,s,l}$ (local flexure of channel lip) as resistance in this case. The above mentioned verifications are to be led for the highest loaded anchor whereas the forces acting on the anchors are to be determined according to Section 4.3.2.

8.4.2 Channel bolt (special screw) and local flexure of channel lip

The characteristic resistances $V_{Rk,s}$ and $M_{Rk,s}$ (failure of special screw) and $V_{Rk,s,l}$ (failure due to local flexure of channel lips) are given in the relevant ETA. In case of a shear force acting with lever arm on the channel bolt Section 5.3.3 is valid.

8.4.3 Concrete pry-out failure

Section 5.3.4 applies. However, $N_{Rk,c}$ shall be calculated according to Equation 8.1. The anchor with the highest ratio $V_{Ed}^{a}/V_{Ed,cp}$ is to be verified.

8.4.4 Concrete edge failure

In case of large distances to the component edge concrete edge break-out does not occur but steel failure of the special screw or channel. For anchor channels with an edge distance in all directions $c \geq 10\ h_{ef}$ and $c \geq 60\ d$ (d = diameter of the special screw), a check of the characteristic concrete edge failure resistance may be omitted. The larger value is decisive.

It is pointed out that in CEN/TS, Part 3, mistakenly the smaller of the above mentioned edge distances is assumed decisive.

The model for the calculation of the characteristic resistance in case of concrete edge failure was suggested by Potthoff (2008). It is based on the following considerations.

Shear loads acting on the anchor channel are transferred by the channel and the anchors in the concrete (see Section 4.3.2). Hence, the concrete edge break-out resistance is influenced by the anchor diameter and the type of channel. In general, for a certain type of channel only one anchor diameter is used. Therefore it is reasonable to combine the influence of channel and anchor on the concrete edge break-out resistance in one factor.

Anchor channels can show any desired number of anchors and shear loads can act at any places vertically to the anchor channel axis. Hence, the model valid for conical concrete break-out under tension load (Section 8.3.4) was transferred to applications with shear loads to calculate the concrete edge break-out resistance.

The characteristic resistance of one anchor loaded perpendicular to the edge corresponds to:

$$V_{Rk,c} = V^0_{Rk,c} \cdot \alpha_{s,V} \cdot \alpha_{c,V} \cdot \alpha_{h,V} \cdot \alpha_{90°,V} \cdot \psi_{re.V} \tag{8.2}$$

The different factors of Equation 8.2 are given below.

8.4.4.1 Characteristic resistance of one anchor (basic resistance)

The basic characteristic resistance of an anchor channel with one anchor loaded perpendicular to the edge not influenced by neighbouring anchors, member thickness or corner effects in cracked concrete is:

$$V^0_{Rk,c} = \alpha_p \cdot \sqrt{f_{ck,cube}} \cdot c_1^{1.5} \quad [\mathrm{N}] \tag{8.2a}$$

with:

α_p factor, given in the relevant ETA. The default value is $\alpha_p \geq 2{,}5$

$f_{ck,cube}$ [N/mm^2], characteristic cube strength (side length 150 mm) of the concrete strength class but noting the limitations given in the relevant ETA

8.4.4.2 Influence of neighbouring anchors

The influence of neighbouring anchors on the concrete edge resistance is taken into account by the factor $\alpha_{s,V}$ according to Equation 8.2b.

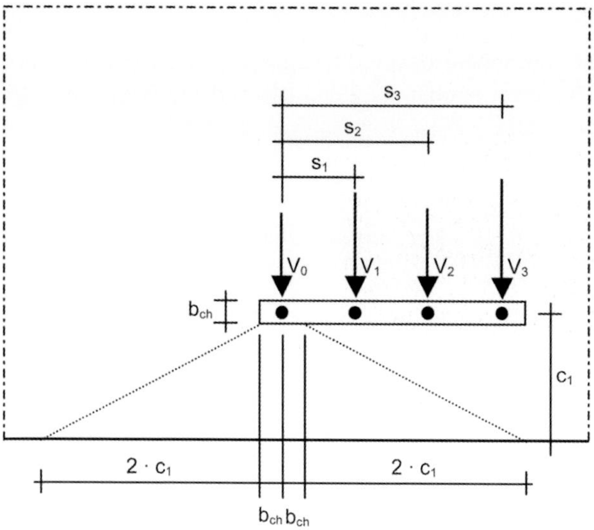

Fig. 8.8 Example of an anchor channel with different shear loads acting vertically to its longitudinal axis

$$\alpha_{s,V} = \frac{1}{1 + \sum_{i=1}^{n} \left[\left(1 - \frac{s_i}{s_{cr,V}}\right)^{1.5} \cdot \frac{V_i}{V_0} \right]} \tag{8.2b}$$

with:

s_i distance between the anchor under consideration and the neighbouring influencing anchor i (Figure 8.8) $\leq s_{cr,V}$

$$s_{cr,V} = 4 \cdot c_1 + 2 \cdot b_{ch} \tag{8.2c}$$

b_{ch} width of anchor channel
V_i shear force of an influencing anchor i
V_0 shear force of the anchor under consideration
n number of anchors within a distance $s_{cr,V}$ to both sides of the anchor under consideration

The factor $\alpha_{s,V}$ substitutes the ratio $A_{c,V}/A^0_{c,V}$ and the factor $\psi_{ec,V}$ in Equation 5.7. Under the assumption of the same characteristic spacing for anchor channels with two anchors the factor $\alpha_{s,V}$ yields practically the same values like the product

$$(A_{c,V}/A^0_{c,V}) \cdot \psi_{ec,V}.$$

Fig. 8.9 Anchor channel with two anchors and a spacing of 5-times the edge distance loaded by a shear load to the edge, after the test (Wohlfahrt (1996))

The characteristic spacing of anchor channels depends not only on the edge distance but also on the width of the channel. The characteristic spacing is bigger compared to post-installed and headed fasteners. The channel detaches early of the concrete at its back and thus represents a disturbance in the concrete structure. Hence, with anchor channels the concrete is higher loaded between the anchors below the rail than with post-installed and headed fasteners having the same spacing. Therefore, the spacing $s_{cr,V}$ to obtain the maximum concrete load-carrying capacity (twice the load-carrying capacity of a single anchor for anchor channels with two anchors) is greater than for post-installed and headed fasteners (Wohlfahrt (1996)). It amounts in case of small edge distance $s_{cr,V} \approx 5 \cdot c_1$, while $s_{cr,V} = 3 \cdot c_1$ is valid for post-installed and headed fasteners. Figure 8.9 shows the concrete break-out of an anchor channel with a distance of the anchor equivalent to 5-times the value of the edge distance. A common break-out body is recognized.

8.4.4.3 Effect of a corner

The influence of a corner on the characteristic edge resistance is taken into account by the factor $\alpha_{c,V}$.

$$\alpha_{c,V} = \left(\frac{c_2}{c_{cr,V}}\right)^{0.5} \leq 1 \tag{8.2d}$$

with:

$$c_{cr,V} = 0.5 \cdot s_{cr,V} = 2 \cdot c_1 + b_{ch} \tag{8.2e}$$

c_2 edge distance (see Figure 8.10a)

If an anchor is influenced by two corners (example see Figure 8.10b), then the factor $\alpha_{c,V}$ according to Equation 8.2d shall be calculated for each corner and the product shall be inserted in Equation 8.2.

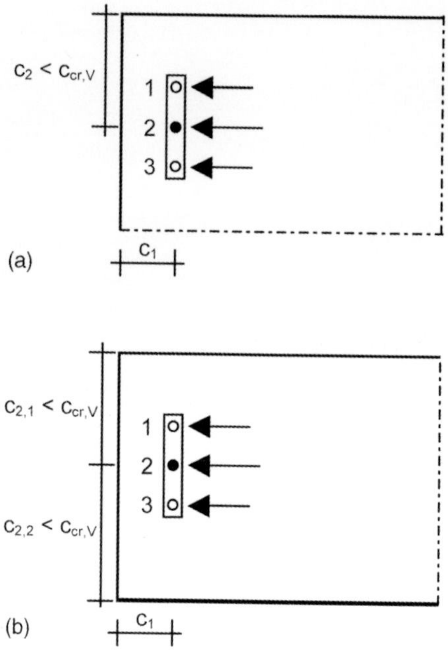

Fig. 8.10 Example of an anchor channel with anchors loaded by shear forces acting vertically to the longitudinal axis of the channel influenced by
(a) one edge
(b) two edges (anchor 2 under consideration)

Results of tests with shear loaded anchor channels in the corner are not present. It is assumed that the factor $\alpha_{c,V}$ according to Equation 8.2d yields conservative results since it was chosen in accordance with the factor $\alpha_{e,N}$ for tension loading.

8.4.4.4 Effect of the thickness of the structural component

The influence of a member thickness $h < h_{cr,V}$ is taken into account by the factor $\alpha_{h,V}$:

$$\alpha_{h,V} = \left(\frac{h}{h_{cr,V}}\right)^{0,5} \leq 1 \tag{8.2f}$$

with:

$$h_{cr,V} = 2 \cdot c_1 + 2 \cdot h_{ch} \tag{8.2g}$$

h_{ch} height of channel bar (Figure 8.11)

The characteristic component thickness for anchor channels depends on the edge distance and the channel bar height. It is bigger compared to post-installed and headed

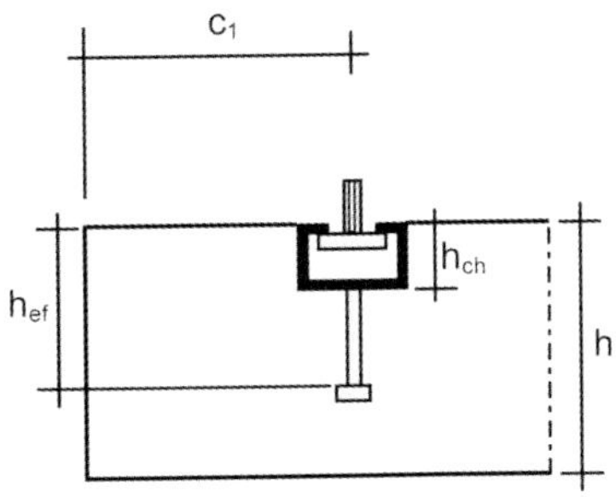

Fig. 8.11 Anchor channel with small edge distance in a thin component

fasteners. It shall be mentioned that in some ETAs the exponent in Equation 8.2f does not amount 0.5 but 2/3.

8.4.4.5 Effect of load parallel to the edge

The factor $\alpha_{90°,V}$ takes into account the influence of shear loads acting perpendicular to the longitudinal axis of the anchor channel and parallel to the edge (see Figure 8.12).

$$\alpha_{90°,V} = 2.5 \tag{8.2h}$$

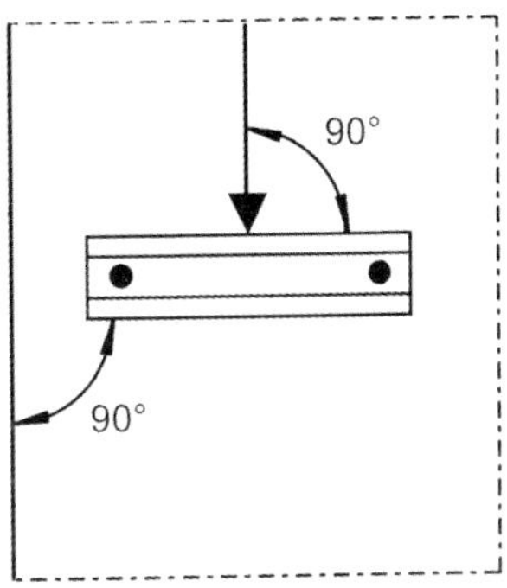

Fig. 8.12 Anchor channel arranged vertically to the component edge, loaded by a shear force acting in parallel with the edge

The factor $\alpha_{90°,V} = 2.5$ agrees with the factor $\psi_{\alpha,V}$ for $\alpha_V = 90°$ (Equation 5.7). If in the example of Figure 8.12 on the remote anchor acts a bigger shear load than on the anchor close to edge both anchors are to be verified. It may be supposed that both anchors do not influence mutually. CEN/TS assumes that anchor channels are inserted only in parallel or vertically to the component edge.

8.4.4.6 Effect of the anchor channel position

Section 5.3.5.7 applies. Nevertheless, the value $\psi_{re,V} > 1$ may be used only in case of applications in the cracked concrete and presence of an edge reinforcement if the height of the anchor channel is $h_{ch} \geq 40$ mm. In other cases of smaller channels and low edge distances the failure crack most probably does not intersect the edge reinforcement.

8.4.4.7 Effect of a narrow member

For an anchor channel in a narrow thin component (Figure 8.13) with $c_{2,\,max.} \leq c_{cr,V}$ ($c_{cr,V}$ according to Equation 8.2e) and $h \leq h_{cr,V}$ ($h_{cr,V}$ according to Equation 8.2g) the calculation after Equation 8.2 leads to conservative results. Besides, $c_{2,max}$ is the maximum of both edge distances parallel to the load direction. More exact results are achieved if the edge distance c_1 is limited to the value c_1' and c_1' is the maximum of the values according to Equation 8.2i$_1$ and 8.2i$_2$.

$$c_1' = (\max\{c_{2,1}; c_{2,2}\} - b_{ch})/2 \tag{8.2i$_1$}$$

$$c_1' = (h - 2 \cdot h_{ch})/2 \tag{8.2i$_2$}$$

The value c_1' is inserted in Equations 8.2a, 8.2c, 8.2e and 8.2g.

Reasons for this approach are to be found in Section 5.3.5.8. It shall be mentioned further that Figure 14 of CEN/TS, Part 3 is not correct because the distance $c_{2,2}$ is measured from the anchor under consideration and not from the neighbouring anchor.

Shear forces acting on fastenings with anchor channels in a narrow thin component should always be taken by a supplementary reinforcement

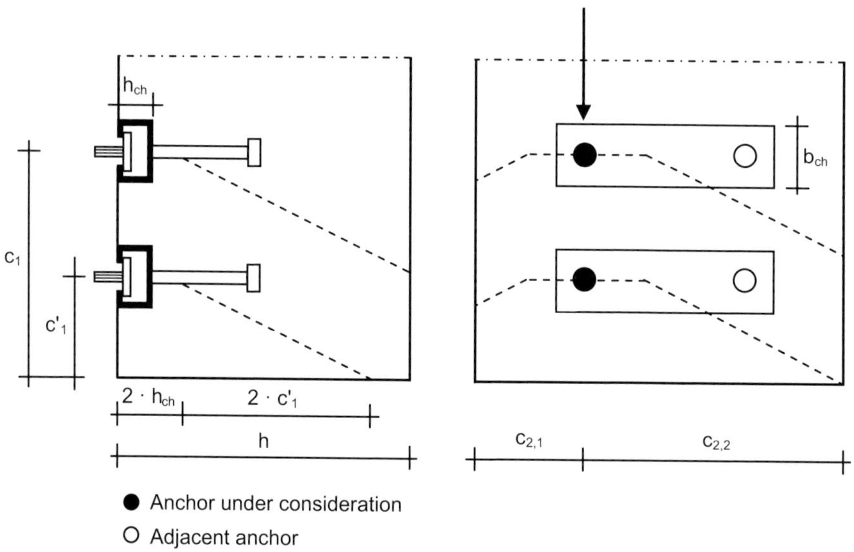

Fig. 8.13 Anchor channel, loaded by a shear load vertically to the longitudinal axis of the channel, influenced by two component edges and the component thickness ($c_{2,2}$ governs on the calculation of c_1')

8.4.4.8 Steel and anchorage failure of the supplementary reinforcement

Sections 7.4.5 and 7.4.6 are valid. In any case loops which enclose the anchors directly are less efficient (see Section 4.4). The assembly of an additional reinforcement as shown in Figure 4.18 is advisable. In general, it should consist of stirrups and straight edge reinforcement. With big edge distances a surface reinforcement with straight bar ends or made of wire mesh can also be used.

According to investigations of Schmid (2010) the concrete is sheared off in front of the anchor channel if there is a high reinforcement ratio of the supplementary reinforcement. This failure mode limits the load-carrying capacity of one anchor of an anchor channel to the value according to Equation 8.3.

$$V_{Rk,c,\max} = \alpha_{re} \cdot c_1^{1,5} \cdot V_{Rk,c} \tag{8.3}$$

with:

α_{re} = 4,2, anchor channels > 28/15
= 2,1, anchor channels of size 28/15

$V_{Rk,c}$ characteristic resistance of concrete edge break-out in uncracked concrete according to Equation 8.2

c_1 edge distance

In CEN/TS the increase of resistance by a supplementary reinforcement is limited by the limitation of the bar diameter to $d_s \leq 16$ mm. This limit is sufficient in many cases. Nevertheless, it is recommended to always perform the verification according to Equation 8.3 if supplementary reinforcement is employed.

It shall be mentioned that the actual ETAs comprise an improved model for the determination of the load-carrying capacity of anchor channels close to the edge with supplementary reinforcement to resist shear loads vertical to the edge which is based on the investigations by Schmid (2010).

8.5 Combined tension and shear loads

The design of anchor channels with and without supplementary reinforcement resisting tension and shear forces is covered by Section 5.4.

Equation 7.8 applies for anchor channels in the component edge and with supplementary reinforcement to resist shear loads. The exponent k_7 amounts $k_7 = 1.0$ (linear interaction) according to investigations by Potthoff (2008).

β_N and β_V shall be determined for all possible failure modes. For reasons mentioned in Section 8.4.1 this is also valid for the failure modes rupture of the anchor or failure of the connection between anchor and channel under shear load.

CEN/TS, Part 3 stipulates to consider failure of the special screw when determining β_N and β_V. This is conservative because the ultimate load of the anchor channel is not influenced by the failure of the special screw. If steel failure is decisive under tension and shear loads, in accordance with CEN/TS, Part 3 the interaction may be performed with the square function (Equation 5.8). This verification is not conservative if the calculated values for concrete failure are only slightly higher than for steel failure. Furthermore the verification for steel failure of the channel with a square function can lead to unsafe results if the load carrying capacity of anchor channels under shear load significantly exceeds tension capacity. This is due to the fact that shear forces acting on the anchor channel induce considerable tension forces in the anchor.

For these reasons after fib (2011) special screw and anchor channel may be verified separately. Equation 5.8 is valid for the special screw. With the anchor channel steel failure of the channel and concrete failure are to be verified separately. Both verifications must be fulfilled and, hence, the more unfavourable case is decisive for the design. Equation 5.8 applies to the verification with steel failure of the anchor channels if the steel load-carrying capacity under shear load is not higher than the value under tension load. Then β_N and β_V are to be calculated for all steel failure modes. The CEN/TS rules described above apply to the verification of the load-carrying capacity of anchor channels without or with supplementary reinforcement with concrete failure. Other details are included in fib (2011). The approach of fib (2011) should be considered for the revision of CEN/TS.

9 Plastic design approach, fastenings with headed fasteners and post-installed fasteners

9.1 General

Forces acting on a fastening can be distributed – as mentioned in Section 4 – to the individual fasteners of a fixture based on the theory of elasticity or theory of plasticity. After the elasticity theory a rigid fixture is assumed and the individual fasteners of the group are loaded to a different level in case of a moment action. However, in case of a design with the theory of plasticity rigid and flexible fixtures are possible. Besides, an essential redistribution of the forces on the individual fasteners of a group occurs after this design procedure, because these fasteners can transmit the loads to neighbouring fasteners after exceeding the elastic limit. This allows for better utilization of the load-carrying capacity of the fasteners compared with the design in accordance with the theory of elasticity. Basic requirements are that

- the fasteners are made of ductile steel and have sufficient elongation capacity and
- the decisive failure mode is always rupture of the steel.

These conditions are hardly to be adhered with actually marketed post-installed fasteners in most applications. However, these conditions can be fulfilled in some applications with headed fasteners. Indeed, the plastic design approach is hardly used in practice, because it is limited to few applications (Section 9.2). Nevertheless, the plastic design approach for the design of post-installed and headed fasteners was included to CEN/TS for reasons of completeness. This approach is described in the following.

The fastener design in accordance with the plasticity theory results in their better utilization compared to the elastic design approach. Indeed, relatively large dimensions of the concrete component serving as anchorage base are necessary (e.g. big edge distances and components with large thickness) to ensure the deep embedment necessary to always employ steel failure of the fastener. These conditions are not found in practice often. With small embedment depth and/or small spacing and edge distances failure by concrete break-out with tension loads or concrete edge break-out with shear loads is decisive, in general. Then the design can only be performed in accordance with the theory of elasticity.

The approach of this section is based on investigations by Cook and Klingner (1989). They examined only fastenings with bending moments acting in one direction. Hence, this design procedure may be used only in this application. For complementary information it is referred to fib (2011).

9.2 Conditions of application

Fasteners may be calculated according to CEN/TS, Part 1, Annex B only with the plastic design approach, if the following conditions are met:

a) Fastening arrangements shown in Figure 9.1 are permitted. Other forms of the attachment are also possible. The fixture may be loaded by normal and shear forces

Design of Fastenings for Use in Concrete – the CEN/TS 1992-4 Provisions. First edition.
Rainer Mallée, Werner Fuchs, Rolf Eligehausen.

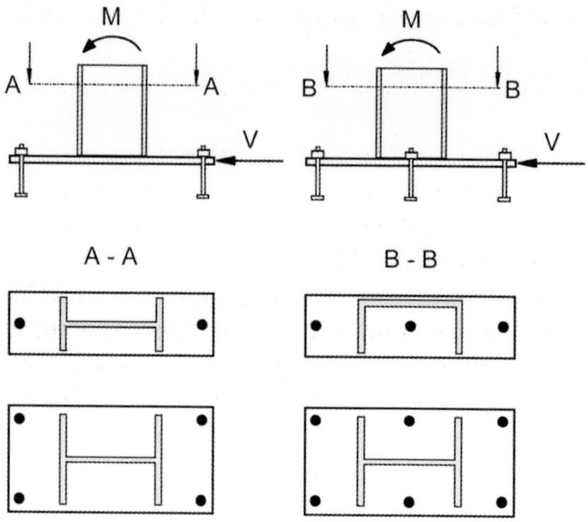

Fig. 9.1 Examples of allowed arrangements of fasteners for which the plastic design approach may be used (Eligehausen and Mallée (2000))

and by a bending moment acting in one direction. The number of fastenings parallel to the axis of bending might be larger than 2. Perpendicular to the axis of bending, in minimum two fasteners have to be always located. Flexible fixtures may be used, if the resultant non-linear load distribution and associated prying forces are taken into account in the calculation of the forces acting on the fasteners.

b) The design resistance of a fastener as governed by concrete failure should exceed the design resistance as governed by steel failure by 25% (see Section 9.4). This ensures ductile behaviour of the fastening.

c) For the same reason the nominal steel strength of the fasteners should not exceed $f_{uk} = 800\ \mathrm{N/mm^2}$, the ratio nominal steel yield strength to nominal ultimate strength shall not exceed $f_{yk}/f_{uk} = 0.8$, and the rupture elongation (measured over a length equal to $5d$) should be at least 12%. The failure mode pull-out is considered non-ductile. Hence, the design value of the pull-out resistance must also be 25% bigger than the design resistance for steel rupture.

d) Special requirements apply to fasteners with reduced cross section. Examples of reduced cross sections are thread areas or in case of bolt anchors necking in the area of the cone. For fasteners loaded in tension, the strength of the reduced section should either be greater than 1,1-times the yield strength of the unreduced section or the stressed length of the reduced section should be at least $5d$ (d = fastening diameter outside reduced section). For fasteners loaded in shear or which shall redistribute shear forces, the start of the reduced section should either be in minimum $5d$ below the concrete surface or in the case of a threaded fastener, the threaded part should extend $2d$ into the concrete. These conditions shall ensure a ductile failure of the fastenings if fasteners with a local reduction of the cross section are used. Requirements concerning the yield strength of the steel as well as the length and the position of the reduced cross section of the fastener shall prevent brittle failure in this area.

e) For fasteners loaded in combined tension and shear, the above conditions shall be met for both load directions.

f) Stand-off installations are not permitted. The steel fixture should be embedded in the concrete or fastened to the concrete surface without an intermediate layer or with a levelling layer of mortar (grout layer). The thickness of the mortar layer with a compressive strength $\geq 30\,N/mm^2$ shall not exceed $d/2$.

g) If the fixture is screwed to the fasteners, the clearance holes of the fixture must be smaller than the values of Table 4.1. Thereby a continuous redistribution of the shear loads to the individual fasteners of the group is ensured. Optimally concerning the load redistribution an optimum is a fastening without hole clearance (welding of headed fasteners to the fixture, a connection without hole clearance usually not to realize in practice or avoidance of the hole clearance by constructive measures like retaining rings or injecting with resin-based mortar).

9.3 Distribution of external forces to the fasteners of a group

CEN/TS, Part 1, Annex B assumes that all fasteners of a group are stressed up to their design steel resistance without taking into account compatibility conditions. This assumption is realistic, because the theory of plasticity presumes a sufficient plastic deformation of every individual fastener of a fastening group and a load redistribution from high loaded to neighbouring fasteners. Moreover, for design purposes, the compressive stress between fixture and concrete may be assumed to be a rectangular concrete stress block. Besides, information is provided on the location of the resultant compressive force depending on the presence of a rigid or a flexible base plate. Figure 9.2 shows examples. Figure 9.2a is valid for a stiff and Figure 9.2b for a flexible base plate. A stiff base plate exists if yielding of the steel is prevented on the edge of the fixture. This condition is satified if Equation 9.1 is fulfilled:

$$M_{yd} > C_{Ed} \cdot a_4 \tag{9.1}$$

with:

M_{yd} design moment that causes yielding of the fixture calculated with $f_{yd} = f_{yk}/\gamma_{Ms}$

γ_{Ms} partial factor; the recommended value is $\gamma_{Ms} = 1.1$

C_{Ed} design resultant compressive force

a_4 distance from the edge of the attached member to the resultant compressive force (see Figure 9.2a)

In case of a flexible base plate, the distance a_5 between the edge of the attached member and the resultant of the compressive reaction of the concrete may be calculated according to Equation 9.2, see Figure 9.2b.

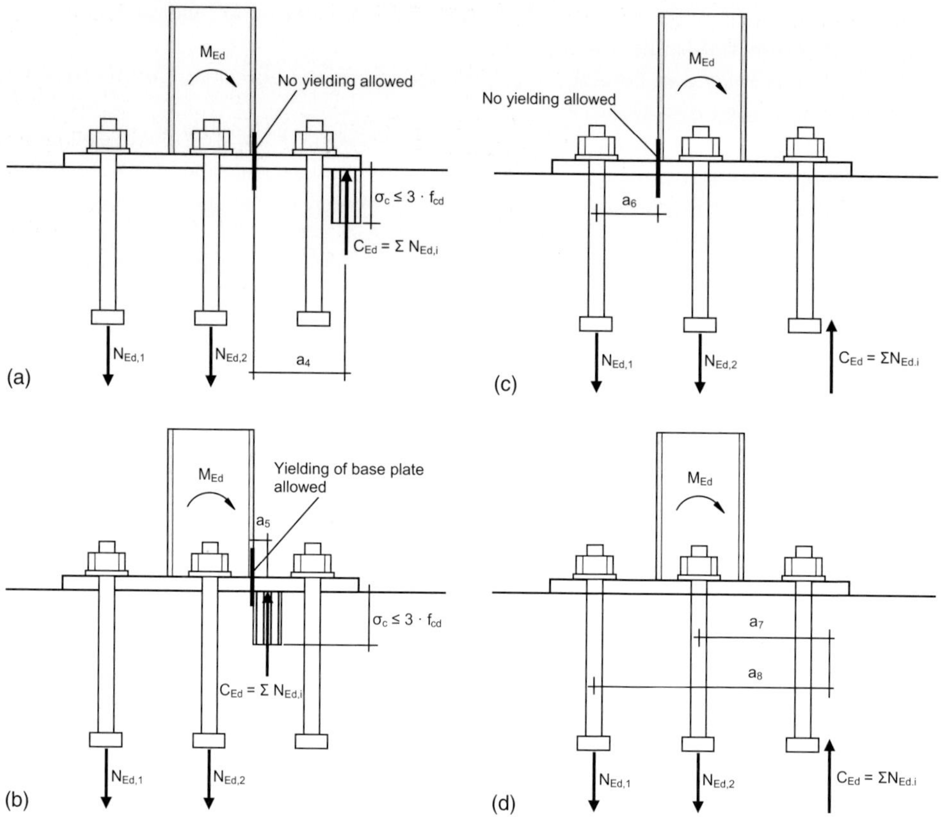

Fig. 9.2 Examples of the location of the resultant concrete compressive force
(a) location of the compression force due to bending in case of a rigid base plate
(b) location of the compression force due to bending in case of a flexible base plate
(c) prevention of the yielding of the fixture on the tensioned side of the connection
(d) condition for fasteners resisting tension loads of about the yield resistance

$$a_5 = M_{yd}/C_{Ed} \tag{9.2}$$

Conservatively, it may be assumed that the compressive reaction is located at either the edge or centroid of the compression element of the attached member.

For both cases (rigid base plate behaviour and flexible base plate behaviour) the formation of a hinge in the base plate on the tension side of the connection shall be prevented. This is satisfied by Equation 9.3 which is valid for one row of fastenings outside the fixture (see Figure 9.2c).

$$M_{yd} > C_{Ed} \cdot a_6 \tag{9.3}$$

with:

C_{Ed} sum of the design tension forces of the outermost row of fastenings

Only those fasteners which satisfy Equation 9.4 shall be assumed to transfer a tension force (see Figure 9.2d).

$$a_7 \geq 0,4 \cdot a_8 \tag{9.4}$$

with:

a_7 (a_8) distance between the resultant compression force and the innermost (outermost) tensioned fastener

9.4 Design of fastenings

In general partial factors used for actions and resistances in the elastic design are also applicable for design based on plastic analysis, except for steel failure. The partial factor for steel $\gamma_{Ms,pl}$ is applied to the yield strength f_{yk} (f_{yk} for self-hardening steel and 0.2% strained length limit for cold-formed steel). The recommended value is $\gamma_{Ms,pl} = 1.2$. However, a CEN-Member State may publish a deviating value in its National Annex.

CEN/TS, Part 1, Annex B, requires verifications for the failure modes steel rupture, pull-out, concrete cone break-out and splitting under tension load as well as steel failure, concrete pry-out and concrete edge break-out under shear load. For the case combined tension load and shear load verification for interaction is to be performed, in addition. The equations corresponding to the verifications of the single failure modes require that the design value of concrete failure is 25% higher than the design value of steel failure. Thus it is ensured that steel failure of all fasteners is to be expected and failure by pull-out, concrete break-out and splitting under tension load or concrete pry-out and concrete edge break-out under shear loads are precluded.

10 Durability

10.1 General

Fasteners shall reliably sustain all assigned actions for the entire working life. This means that the design working life of the fastener shall not be less than that of the fixture.

Structural corrosion protection measures such as use of corrosion-insensitive materials, low as structured connections, prevention of pollutant sediments from the air, drainage of condensed water, sufficient ventilation are already to be considered in the design of the connection and the choice of the fasteners.

CEN/TS is based on an assumed intended working life of the fastener of at least 50 years. Informative Annex C "Durability" includes information on how to ensure the intended working life with respect to corrosion aspects.

CEN/TS rules for durability of fasteners are based on the exposure classes known by EN 1992-1-1:2004 (Eurocode 2). These are summarized user-friendly in the following environmental conditions:

- dry, internal conditions
- external atmospheric or permanently damp internal exposure
- high corrosion exposure by chloride and sulphur dioxide.

The industry provides for these areas of application fasteners with appropriate coatings or of suitable materials. If different metals are used for fastener and fixture, electrolytic corrosion must be prevented by suitable isolating interlayers or by the choice of compatible materials.

10.2 Fasteners in dry, internal conditions

The dry, internal conditions are similar to exposure class XC1 according to EN 1992-1-1:2004 (Eurocode 2).

In general, no special corrosion protection is necessary for fasteners. Fasteners made of carbon steel are normally electro galvanized (zinc thickness 5 μm). This type of coating provided for preventing corrosion during storage prior to use is considered sufficient to ensure in the structure a working life of 50 years. Malleable cast iron parts which are hardly used in fastening practice, in general, do not require any additional protection.

10.3 Fasteners in external atmospheric or in permanently damp internal exposure and high corrosion exposure

Table 10.1 demonstrates that for dry indoor conditions the normal electro galvanizing cannot ensure the load-carrying capacity and serviceability of a fastener for the necessary service life of 50 years even under low corrosive exposure. This applies in particular to buildings where the fastenings are not accessible.

Design of Fastenings for Use in Concrete – the CEN/TS 1992-4 Provisions. First edition.
Rainer Mallée, Werner Fuchs, Rolf Eligehausen.

Table 10.1 Average rates of zinc plating loss in Germany (after Beratung Feuerverzinken (Consulting hot-dip galvanizing) (1983) and DIN EN ISO 12944-2 (1998))

Atmosphere	Loss [μm/Jahr]	
	Beratung Feuerverzinken (1983) (Consulting hot-dip galvanizing)	DIN EN ISO 12944-2 (1998)
Rural	1.3–2.5	0.1–0.7
Urban	1.9–2.5	0.7–2.1
Industrial	6.4–13.8	4.2–8.4
Marine	2.2–7.2	2.1–4.2

Fasteners of stainless steels ensure a sufficient corrosion resistance under the following corrosive environments:

- spaces with permanent high level of humidity or outdoor conditions
- coastal regions with high salt concentration of the air
- splash zone of seawater
- industrial zones with high level of air pollution
- buildings for infrastructure with high level of air pollution, for example, multi-story car parks
- splash zones of roads (de-icing salt)
- structures with permanently active chlorine vapours, for example, public swimming-pools
- structures heavily loaded with ammonium vapours, for example cattle stables
- structures with tannic acid-containing timber (e.g. oak).

CEN/TS gives the following instructions to the materials to be used.

10.3.1 Fastenings in external atmospheric or in permanently damp internal exposure

These conditions (e.g. fastenings in spaces with a permanent high level of humidity and external atmospheric, in marine climate or industrial atmosphere etc.) are similar to exposures XC2, XC3 and XC4 according to EN 1992-1-1:2004 (Eurocode 2).

Normally stainless steel fasteners of appropriate grade should be used, for example austenitic steels with at least 17 to 18% chromium and 12 to 13% nickel and addition of molybdenum for example material 1.4401, 1.4404, 1.4571, 1.4578 and 1.4439 according to EN 10088-2 or EN 10088-3. Equivalent types of steel or materials in accordance with existing national rules may be also employed.

10.3.2 Fasteners in high corrosion exposure by chloride and sulphur dioxide

Examples for these conditions are permanent, alternating immersion in seawater or the splash zone of seawater, chloride atmosphere of indoor swimming pools or atmosphere with extreme chemical pollution (e.g. in desulphurization plants or road tunnels, where de-icing materials are used. These conditions are similar to exposure classes XD and XS according to EN 1992-1-1:2004 (Eurocode 2).

The metal parts of the fastener should be made of a stainless steel suitable for the high corrosion exposure and shall be in accordance with national rules. In general stainless steel with about 20% chromium, 20% nickel and 6% molybdenum for example so-called HCR-materials (HCR high corrosion resistance) such as 1.4565, 1.4529 und 1.4547 according to EN 10088-2 or EN 10088-3 or equivalent should be used under high corrosion exposure.

11 Exposure to fire

11.1 General

CEN/TS, informative Appendix D addresses the design of fastenings subjected to fire exposure. With the increasing use of post-installed fasteners in structural engineering the subject of fire resistance has grown in relevance. The design procedure is based on a proposal in European Organization for Technical Approvals (EOTA) (2004).

Fasteners installed in concrete are not only used to connect load bearing structural components but also attachments for non-structural components such as air conditioning systems, suspended ceilings, electrical equipment, and piping systems. Some of these components, such as fire sprinkler piping or fire detector installation, may be decisive to the fire protection of the structure. The premature failure of these components may pose a falling hazard to occupants and first responders could block escape routes and delay or even preclude the efforts of the fire brigades.

CEN/TS provides a universally valid conservative approach for the design of cast-in headed fasteners, expansion fasteners made of steel, undercut fasteners and concrete screws in the fire case.

Higher values of fire resistance than predicted by CEN/TS are possible, if the fastener manufacturers perform fire tests according to the regime given in European Organization for Technical Approvals (EOTA) (2004) and these product specific values are stated in an ETA.

For post-installed chemical fasteners the fire resistance in the cases of combined bond and concrete failure is product dependent. Hence, the fire resistance cannot be determined according to CEN/TS. Anchor channels are not covered. Here the statements of the approval documents or the information of the manufacturers are to be observed.

The design procedure is based on the following restrictions:

- the fire resistance of fastener and fixture is adequate.
- the fire resistance is classified according to EN 13501-2. It is based on the standard fire time-temperature curve of ISO 834 or DIN 4102, which is intended to simulate burning timber (Figure 11.1).
- the design method covers fasteners with a fire exposure from one side only. For fire exposure from more than one side, the design method may be used only, if the edge distance of the fastener is $c \geq 300\,\text{mm}$ and $c \geq 2 \cdot h_{ef}$.
- in other applications such as the petrochemical industry, the time-temperature curve is steeper since the fuel of the fire (hydrocarbon instead of wood) burns faster. Where these special conditions apply other curves, such as the UL 1709 hydrocarbon curve, are used. Special time-temperature curves have also been developed for tunnel fires based on the experience from the Mont Blanc and Gotthard tunnel fires. As a consequence, for the above mentioned applications the design procedure given in CEN/TS is not applicable.

Design of Fastenings for Use in Concrete – the CEN/TS 1992-4 Provisions. First edition.
Rainer Mallée, Werner Fuchs, Rolf Eligehausen.

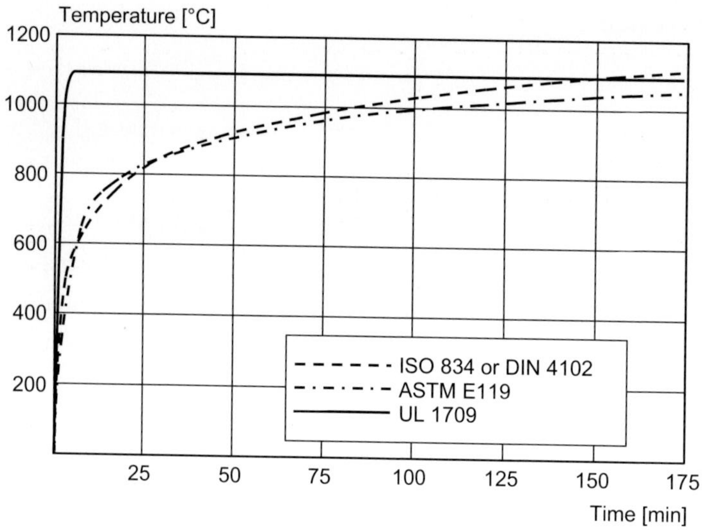

Fig. 11.1 Standard time-temperature curves according to different regulations

11.2 Basis of design

The results of tests (Reick (2001)) also indicated that the same failure mode types as occur in testing at service temperature are observed in testing under fire exposure. These same failure modes must also be verified in the fire design.

In fire exposure fasteners might fail by means of steel failure because the strength of steel decreases significantly with rising temperature. Steel failure is often observed in fire testing. Failure is characterized by rupture of the anchor rod or stripping of the threads. The resistance depends on the duration of the fire, the type of steel and the diameter of the fastener. Stainless steels have a longer resistance to fire than zinc-plated carbon steels (Figure 11.2). For the same steel stress, fasteners with smaller diameters fail sooner than fasteners with large diameters (Figure 11.3).

Concrete break-out occurs preferably with fastenings with small embedment depth or small spacing. In case of fire exposure the tensile strength of the concrete is significantly reduced. Furthermore the higher temperature gradients between the fire-exposed surface and the lower lying layers cause thermal tension possibly resulting in large cracks inside the component. In addition, free water bound physically in the concrete vaporizes and generates additional concrete stress often resulting in surface spalling. This effect can be minimized if the concrete component is designed according to EN 1992-1 (2004). The concrete should be produced with quartzite additives and the concrete member must be protected from direct moisture exposure.

The pull-out failure of mechanical expansion fasteners is determined by the combination of materials used for the fastener in the anchorage zone, the coatings on the cone and the expansion sleeve as well as its geometry.

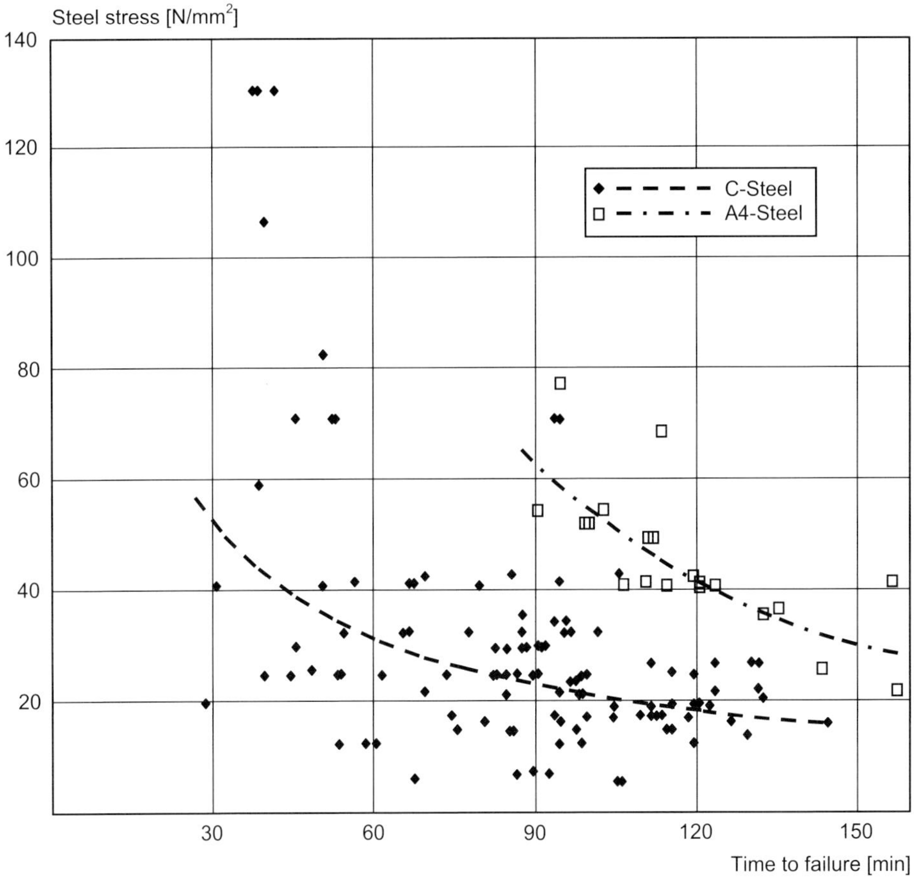

Fig. 11.2 Ultimate strength of carbon steel and stainless steel as a function of time until failure (Reick (2001))

The verification of splitting failure due to loading under fire exposure is not required because the splitting forces are assumed to be taken up by the reinforcement

At the ultimate limit state of the anchor under fire exposure the procedure of Section 3.2 for service temperature applies accordingly. For all load directions and failure modes it shall be shown that:

$$E_{d,fi} \leq R_{d,fi} \tag{11.1}$$

with:

$E_{d,fi}$ design value of action to be resisted under fire exposure

$$= E_{k,fi} \cdot \gamma_{F,fi} \tag{11.2}$$

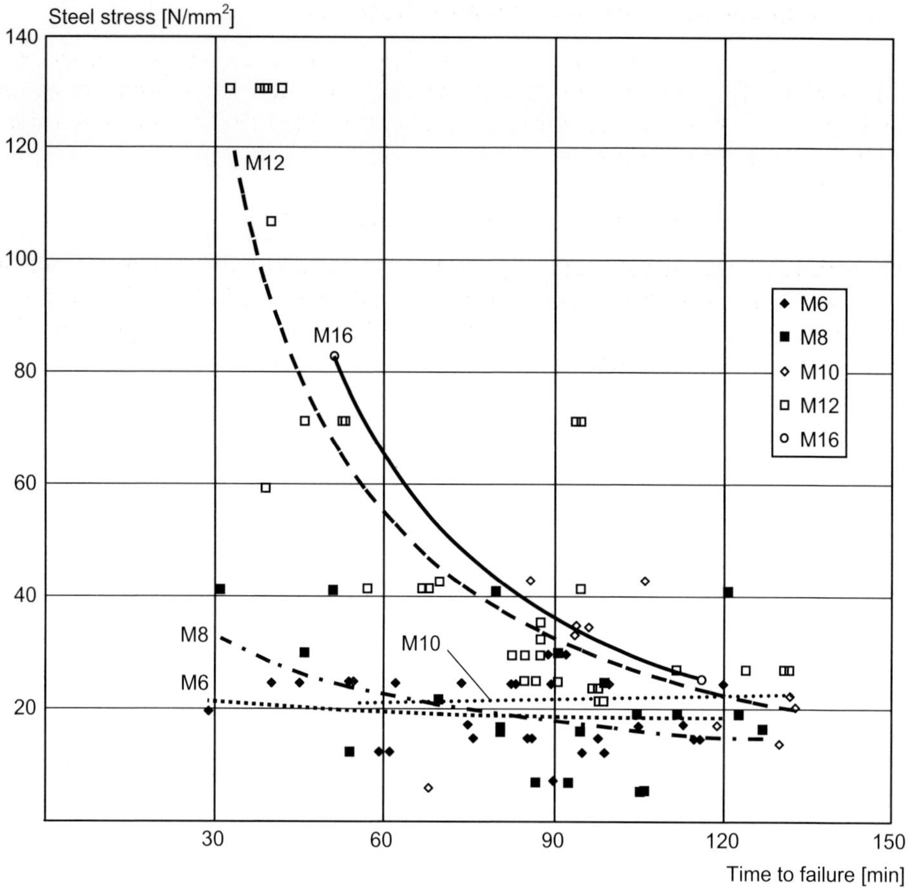

Fig. 11.3 Ultimate strength of carbon steel for diameter M6 to M16 as a function of time until failure (Reick (2001))

$E_{k,fi}$ characteristic value of actions to be resisted under fire exposure
$\gamma_{F,fi}$ partial factor for actions during fire exposure
$R_{d,fi}$ design value of resistance corresponding to fire exposure

$$= R_{k,fi}/\gamma_{M,fi} \qquad (11.3)$$

$R_{k,fi}$ characteristic value of resistance corresponding to fire exposure
$\gamma_{M,fi}$ partial (safety) factor for resistance corresponding to fire exposure

The values for the partial factors depend on the European country and might be determined by a National Annex to CEN/TS. In general, the values are set to $\gamma_{F,fi} = 1.0$ for actions and $\gamma_{M,fi} = 1{,}0$ for materials.

11.3 Resistances under tension and shear load

The CEN/TS design procedure applies if for a fastener no deviating results from tests are present which can be listed for example in an ETA. The design approach is based on characteristic values of the fire resistance which are verified by numerous test results with different fasteners under fire exposure.

11.3.1 Steel failure under tension load and shear load

The characteristic tension strength $\sigma_{Rk,s,fi}$ is given in Table 11.1 for carbon steel and in Table 11.2 for stainless steel. These values are valid under tension and shear loading since limited number of tests have indicated, that the ratio of shear strength to tensile strength increases to approximately 1.0 under fire conditions, in contrast to the behaviour at service temperature levels where the ratio is in the order of 0.6. Then for steel failure the design value of resistance corresponding to fire exposure is:

$$R_{d,fi,s} = \sigma_{Rk,s,fi} \cdot A_s / \gamma_{M,s,fi} \quad (11.4)$$

Table 11.1 Characteristic tension strength of a carbon steel fastener under fire exposure

Anchor bolt/thread diameter	Embedment depth h_{ef} [mm]	Characteristic tension strength of a fastener made of carbon steel $\sigma_{Rk,s,fi}$ [N/mm^2]			
		30 min	60 min	90 min	120 min
Ø 6/M6	≥30	10	9	7	5
Ø 8/M8	≥30	10	9	7	5
Ø 10/M10	≥40	15	13	10	8
Ø 12/M12 and greater	≥50	20	15	13	10

Table 11.2 Characteristic tension strength of a stainless steel fastener under fire exposure

Anchor bolt/thread diameter	Embedment depth h_{ef} [mm]	Characteristic tension strength of a fastener made of stainless steel $\sigma_{Rk,s,fi}$ [N/mm^2]			
		30 min	60 min	90 min	120 min
Ø 6/M6	≥30	10	9	7	5
Ø 8/M8	≥30	20	16	12	10
Ø 10/M10	≥40	25	20	16	14
Ø 12/M12 and greater	≥50	30	25	20	16

with:

$\sigma_{Rk,s,fi}$ according to Tables 11.1 or 11.2
$A_{s,fi}$ net cross section [mm^2]

11.3.2 Steel failure under shear load with lever arm

The characteristic bending resistance $M^0_{Rk,s,fi}$ is:

$$M^0_{Rk,s,fi} = 1.2 \cdot W_{el} \cdot \sigma_{Rk,s,fi} \tag{11.5}$$

with:

$\sigma_{Rk,s,fi}$ according to Tables 11.1 or 11.2
W_{el} elastic section modulus of the governing section

11.3.3 Pull-out under tension load

The pull-out resistance under tension load is calculated as one quarter of the strength determined according to the relevant ETA for use in cracked concrete C20/25 at ambient temperature under service conditions for fire exposures up to 90 minutes (R90); for fire exposure up to 120 minutes (R120) the reduction is one fifth:

$$\text{R90:} \quad N_{Rd,p,fi(90)} = 0.25 \cdot N_{Rk,p}/\gamma_{Mp,fi} \tag{11.6a}$$

$$\text{R120:} \quad N_{Rd,p,fi(120)} = 0.20 \cdot N_{Rk,p}/\gamma_{Mp,fi} \tag{11.6b}$$

This design procedure is valid only for post-installed fasteners with an ETA for applications in the cracked concrete because the fire resistance in case of pull-out failure is derived from the resistance at ambient temperature in the cracked concrete.

11.3.4 Concrete break-out under tension load and concrete pry-out failure under shear load

Concrete break-out resistance under fire exposure is similarly calculated as a percentage of the resistance at ambient temperature but as a function of the embedment depth if the embedment depth is $h_{ef} \leq 200$ mm. For fire exposure up to 90 minutes (R90) the resistance load is calculated as 25% of the strength determined according to the relevant ETA for use in cracked concrete C20/25 at ambient temperature independent of the load direction; for fire exposure up to 120 minutes (R120) the resistance is 20%:

Tension load:

$$\text{R90:} \quad N^0_{Rk,c,fi(90)} = \frac{h_{ef}}{200} \cdot N^0_{Rk,c} \leq N^0_{Rk,c} \tag{11.7a}$$

$$\text{R120:} \quad N^0_{Rk,c,fi(120)} = 0,8 \cdot \frac{h_{ef}}{200} \cdot N^0_{Rk,c} \leq N^0_{Rk,c} \tag{11.7b}$$

Shear load and pry-out failure:

R90: $$V_{Rk,cp,fi(90)} = k \cdot N_{Rk,c,fi(90)} \quad (11.8a)$$

R120: $$V_{Rk,cp,fi(120)} = k \cdot N_{Rk,c,fi(120)} \quad (11.8b)$$

with:

k factor to be taken from the relevant ETA (ambient temperature)

11.3.5 Concrete edge failure under shear load

For fire exposures up to 90 minutes (R90) the characteristic concrete edge resistance under shear load is calculated as 25% of the strength for use in cracked concrete C20/25 at ambient temperature; for fire exposure up to 120 minutes (R120) the resistance is 20%:

R90: $$VRk, c, fi(90) = 0.25 \cdot V_{Rk,c} \quad (11.9a)$$

R120: $$V_{Rk,c,fi(120)} = 0.20 \cdot V_{Rk,c} \quad (11.9b)$$

with:

$V_{Rk,c}$ characteristic resistance for cracked concrete C20/25 under ambient temperature

The interaction conditions at ambient temperature according to the relevant product specific part of CEN/TS may be taken with the characteristic resistances under fire exposure for the different loading directions for combined tension and shear loads.

12 Seismic loading

12.1 General

CEN/TS, Part 1, Section 8 provides design requirements for fasteners under seismic loading. Informative Annex E gives information on the derivation of loads acting on the fasteners. These requirements are necessary additions to EN 1998-1:2004 (Eurocode 8), Section 4.3.5. While EN 1998–1 provides requirements for the design of non-structural elements in this Section, it ignores the vertical accelerations in the calculation of actions. This could lead to unsafe designs for fastenings securing non-structural items. If for example a water supply line for fire fighting is fixed to a reinforced concrete slab, vertical accelerations would result in a significant increase of the loads acting on the fastening. If a water supply line is situated on a corbel fastened to a wall, vertical accelerations lead not only to an increase of the shear load, but increase also the actions on the upper post-installed fasteners of the fastening of the corbel by the developing additional moment. For these reasons CEN/TS, Annex E defines additional requirements for fasteners under seismic excitations.

Annex E provides a pragmatic approach to establish the dynamic characteristics of the components. This approach is based on assumptions likely satisfactory for most cases. However, CEN/TS points out that the designer is responsible to see that the requirements EN 1998-1:2004 (Eurocode 8) are fulfilled. Moreover, it is pointed out that Annex E might serve as a basis for the determination of loads acting on the fasteners in case of an earthquake. The approach given in Annex E does not necessarily assure operability of the non-structural element during or after an earthquake. In this case verifications specific to the application are to be performed.

12.2 Additions and alterations to EN 1998-1:2004 (Eurocode 8)

If non-structural elements are fastened on the floor, the friction between component and concrete as a result of the gravity load has a beneficial effect in theory. This frictional force, however, should be ignored according to CEN/TS because as a result of horizontal accelerations during an earthquake tilting movements of the component up to lifting of the component might occur. Then the frictional forces having an effect in the static case are not effective any more.

The horizontal effects of the seismic action may be determined by applying a horizontal force F_a to the centre of gravity of the non-structural element acting in the most unfavourable direction which is defined as follows:

$$F_a = (S_a \cdot W_a \cdot \gamma_a)/q_a \tag{12.1}$$

with:

S_a horizontal seismic coefficient applicable to non-structural elements
W_a weight of the element to be fastened

Design of Fastenings for Use in Concrete – the CEN/TS 1992-4 Provisions. First edition.
Rainer Mallée, Werner Fuchs, Rolf Eligehausen.

γ_a importance factor of the element, to be determined according to EN 1998-1:2004, Section 4.3.5.3, however, according to CEN/TS, Annex E, Section E.4.1 required to be 1.5 in minimum

q_a behaviour factor of the element (Annex E, Table E.1)

The horizontal seismic coefficient S_a may be calculated after Equation 12.2:

$$S_a = \alpha \cdot S \cdot \left[\left(1 + \frac{z}{h}\right) \cdot A_a - 0,5\right] \geq \alpha \cdot S \tag{12.2}$$

with:

α ratio of the design ground acceleration on type A ground, a_g to the acceleration of gravity g

S soil factor

z height of the non-structural element above the level of application of the seismic action

h building height, measured from the foundation or from the top of a rigid basement

$$A_a = \frac{3}{1 + \left(1 - \frac{T_a}{T_1}\right)^2} \tag{12.3}$$

T_a fundamental vibration period of the non-structural element

T_l fundamental vibration period of the building in the relevant direction

If the values of T_a and T_1 are not known, the values in Annex E, Table E.1 may be used.

The vertical effects of the seismic action may be determined by applying to the centre of gravity of the non-structural element a vertical force F_{va} which is defined as follows:

$$F_{va} = (S_{va} \cdot W_a \cdot \gamma_a)/q_a \tag{12.4}$$

with:

$$S_{va} = \alpha_V \cdot A_a \tag{12.5}$$

α_V ratio of the vertical design ground acceleration on type A ground, a_{vg} to the acceleration of gravity g

A_a response amplification factor

The factors W_a, γ_a and q_a are explained in Equation 12.1. Values of the response amplification factor A_a and behaviour factor q_a for non-structural elements are given in Annex E, Table E.1. Table E.1 distinguishes between architectural elements (exterior wall elements, partitions, canopies etc.), mechanical equipment (storage vessels and water heaters, piping, fire suppression piping etc.), electrical equipment (electrical communication equipment, light fixtures etc.), interior equipment (storage racks, computer access floors, elevators etc.) and other unspecified equipment. Detailed information is given in Annex E, Table E.1.

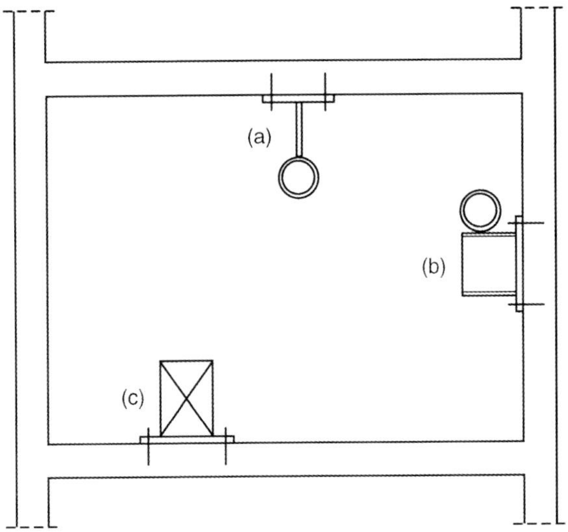

Fig. 12.1 Consideration of the vertical acceleration of the earthquake effect
(a) fixing in a ceiling, vertical acceleration F_{va} shall be considered
(b) fixing to a wall, vertical acceleration F_{va} shall be considered
(c) fixing on the ground slab where vertical acceleration F_{va} might be neglected

Figure 12.1 indicates in which cases the vertical effects of the seismic action may be considered or neglected.

12.3 Verification of seismic loading

12.3.1 General

This section provides requirements for fastenings used to transmit seismic actions (tension loads, shear loads and combined tension and shear loads) when connecting structural elements or anchoring non-structural elements to structural components.

CEN/TS does not cover fixtures with a grout layer $>d/2$ (Figure 4.12) since the behaviour of fastenings for this kind of application is hardly investigated.

Fasteners used to resist seismic actions shall meet all applicable requirements for non-seismic applications.

Only fasteners qualified for seismic applications shall be used. A European guideline providing the relevant prequalification tests and requirements for the assessment is under preparation by EOTA. This guideline will be published in the first half of 2013.

The concrete in the region of the fastening shall be assumed to be cracked when determining the design resistance.

The CEN/TS provisions do not apply to the design of fastenings in critical regions of concrete structures, where concrete spalling or yielding of the reinforcement with wide cracks might occur, for example in plastic hinge zones (critical regions) of reinforced concrete structures during the seismic event. In these regions the crack widths might be much larger than the crack width in the specimen used for the fastener prequalification. The length of the critical region is defined in EN 1998-1:2004 (Eurocode 8).

Displacement of the fastening shall be accounted for in the design based on engineering judgment. This is especially valid when fixing for example structural elements, non-structural elements of great importance or of particularly dangerous nature or when the structure shall demonstrate no essential displacements and be operable after the seismic event. The fastener displacements for seismic applications will be provided in the future ETAs of the fasteners.

Determination of distribution of forces to the individual fasteners of a group shall take into account the stiffness of the fixture and its ability to redistribute loads to other fasteners in the group beyond yield of the fixture.

Annular gaps between a fastener and its fixture should be avoided for seismic design situations to establish a uniform distribution of the shear load to the individual fasteners of a group. Where in minor non-critical applications this requirement is not fulfilled, the effect of the annular gap ($d_f \leq d_{f,1}$) on the distribution of shear loads in the case of groups and on the resistance should be taken into account.

Loosening of the nut or screw shall be prevented by appropriate measures.

12.3.2 Derivation of actions

The design value of the effect of seismic actions E_d acting on the fixture shall be determined according to EN 1998-1:2004 (Eurocode 8). The requirements for non-structural elements given in Section 12.2 shall be observed.

12.3.3 Resistance

The partial factor γ_M shall be determined according to Section 3.3. The seismic design resistance $R_{d,eq}$ of a fastening under seismic actions is calculated according to Equation 12.6:

$$R_{d,eq} = \alpha_{eq} \cdot \frac{R_{k,eq}}{\gamma_M} \tag{12.6}$$

with:

α_{eq} = 0.75 for concrete related failures: concrete cone, pull-out, blow-out and splitting failure under tension loading; pry-out and concrete edge failure under shear loading
= 1.0 for steel failure

$R_{k,eq}$ characteristic seismic resistance for a given failure mode

The characteristic seismic resistances are calculated according to Sections 5 to 7 for the failure modes concrete cone break-out and blow-out in case of tension loading and concrete pry-out and concrete edge failure in case of shear loads. In general, splitting of the concrete element does not have to be verified because the splitting forces are resisted by the reinforcement and the design is performed assuming cracked concrete. The characteristic seismic resistances for steel failure under tension and shear loading as well as the characteristic seismic pull-out resistances under tension load are given in the future ETAs.

The resistances for concrete failure are calculated as in normal persistent and transient design situations. In case of seismic events, however, wider cracks compared to persistent and transient design situations are to be expected. In addition the displacements of the fasteners indicate a higher scatter resulting in a more non-uniform resistance of the individual fasteners in a group. Both effects lead to a reduction of the ultimate resistance. This is considered by the factor α_{eq}.

When the fastening design includes seismic actions one of the following conditions shall be satisfied:

a) The anchorage is either designed for the minimum of the force corresponding to yield of a ductile steel component taking into account over-strength (Figure 12.2(a) and (b)) or the maximum force that can be transferred to the connection by the attached component or structural system (Figure 12.2(c)).
b) The fastener is designed for ductile steel failure (Figure 12.3). To ensure ductile steel failure Equation 12.7 and the requirements of Sections 9.2 c), d) und e) shall be fulfilled:

$$R_{k,s,eq} \leq 0.6 \cdot \frac{R_{k,conc,eq}}{\gamma_{inst}} \tag{12.7}$$

with:

$R_{k,s,eq}$ characteristic seismic resistance for steel failure
$R_{k,conc,eq}$ characteristic seismic resistance for all non-steel failure modes such as concrete cone, blow-out or pull-out under tension loading or pry-out or concrete edge failure under shear loading
γ_{inst} partial factor for installation safety (see Section 3.3.3.1), given in the relevant ETA

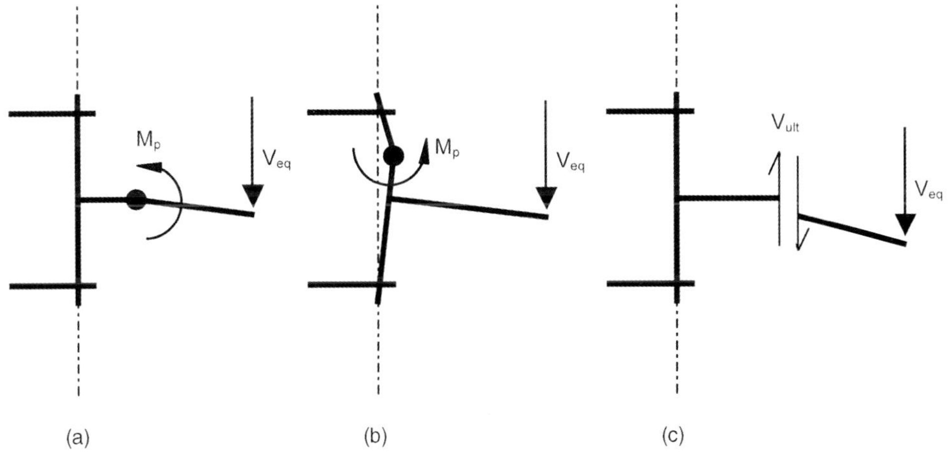

Fig. 12.2 Protection of the fastener under seismic actions
(a) yielding in the fixture
(b) yielding of the base plate
(c) design for the biggest force which can be transferred to the fastener by the fixture or structure

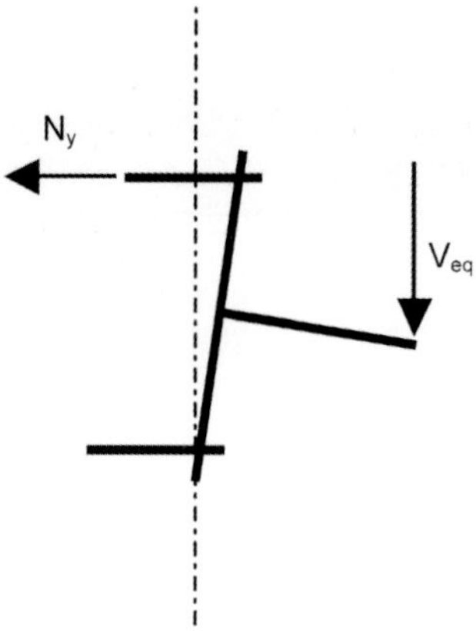

Fig. 12.3 Yielding of the ductile fastener

c) For non-structural elements, brittle failure of the fastening may be permissible only if the seismic design resistance after Equation 12.6 is taken as at least 2.5 times the effect of the applied seismic action E_d of the attached non-structural element (see Equation 12.8). For structural elements, brittle failure of the fastening is not allowed.

$$2.5 \cdot E_d \leq \alpha_{eq} \cdot \frac{R_{k,eq}}{\gamma_M} \tag{12.8}$$

with:

α_{eq} see Equation 12.6

After condition (a) the fastening is designed for the maximum load to be resisted by the fixture or attached component. The fastening is protected against overloading. In this case, hence, brittle concrete failure is permissible.

Condition (b) ensures ductile steel failure of the fastening and excludes brittle concrete failure with high probability.

Minimum spacing and edge distance for persistent and transient design situations apply also in case of seismic actions, in general.

The interaction between tension and shear forces shall be determined assuming a linear interaction as given in Equation 12.9, unless different product specific interaction relations for seismic applications are provided in the relevant ETA.

$$\frac{N_{Ed,eq}}{N_{Rd,eq}} + \frac{V_{Ed,eq}}{V_{Rd,eq}} \leq 1 \tag{12.9}$$

In Equation 12.9 the largest ratios $N_{Ed,eq}/N_{Rd,eq}$ and $V_{Ed,eq}/V_{Rd,eq}$ for the different failure modes shall be inserted.

13 Outlook

CEN/TS 1992-4 is a pre-standard (Technical Specification, TS) with an initial validity of three years. In early summer 2012 committee CEN/TC 250/SC 2 has decided to transform this CEN/TS into Part 4 of Eurocode 2. This new EN 1992-4 will be editorially a completely restructured version of the actual CEN/TS. It will be published as one document most presumably in 2014.

Design of Fastenings for Use in Concrete – the CEN/TS 1992-4 Provisions. First edition.
Rainer Mallée, Werner Fuchs, Rolf Eligehausen.

References

Appl, J.J. (2009) Tragverhalten von Verbunddübeln unter Zugbelastung (Load-bearing behaviour of adhesive anchors under tension loading). Doctor thesis, Universität Stuttgart, 2009 (in German).

Beratung Feuerverzinken (Consulting hot-dip galvanizing) (1983) Zink statt Rost (Zinc instead of corrosion). Hagen (in German).

CEN/TS 1992-4 (2009) *Design of Fastenings for Use in Concrete*, Edition 2009.

Comité Euro-International du Béton (CEB) (1995) Design of Fastenings in Concrete, CEB Bulletin 226, Thomas Telford Services Ltd., Lausanne, 1997. p. 1–144.

Cook, R. and Klingner, R. E. (1989) Behaviour and design of ductile multiple anchor steel to concrete connections. Report No. CTR 1126-3, University of Texas at Austin.

Cook, R.A., Kunz, J., Fuchs, W., and Konz, R.C. (1998) Behaviour and design of single adhesive anchors under tensile load in uncracked concrete. *ACI Structural Journal*, **95**, 9–26.

Deutsches Institut für Bautechnik (1993) Bemessungsverfahren für Dübel zur Verankerung in Beton (Anhang zum Zulassungsbescheid) (Design method for anchorage to concrete (Annex for approval certificates)). Edition Juni 1993, Berlin (in German).

Deutsche Institut für Normung (DIN) DIN EN ISO 12944-2 (1998) Beschichtungsstoffe – Korrosionsschutz von Stahlbauten durch Beschichtungssysteme(Paints and varnishes – Corrosion protection of steel structures by protective paint systems – Part 2: Classification of environments (ISO 12944-2:1998); German version EN ISO 12944-2:1998.

Deutsche Institut für Normung (DIN) DIN 4102 (1998) Brandverhalten von Baustoffen und Bauteilen (Technical building regulations forfir behaviour of building materials, building componenets and special components), Edition October 1998 (in Geman).

Deutsches Institut für Normung (DIN) DIN SPEC 1021-4 (2009): Bemessung der Verankerung von Befestigungen in Beton. Berlin (in Geman).

Eligehausen, R., Mallée, R., and Rehm, G. (1984) Befestigungen mit Verbundankern (Fixing with bonded anchors). *Betonwerk + Fertigteil-Technik*, 686–692, 781–785 and 825–829 (in German).

Eligehausen, R., Fuchs, W., Lotze, D., and Reuter, M. (1989) Befestigungen in der Betonzugzone (Fixings in the concrete tensile zone). *Beton- und Stahlbetonbau*, **84**, 27–32 and 71–74 (in German).

Eligehausen, R., Fuchs, W., Ick, U., Mallée, R., Reuter, M., Schimmelpfennig, K., and Schmal, B. (1992) Tragverhalten von Kopfbolzenverankerungen bei zentrischer Zugbeanspruchung (Behaviour of headed anchors under concentrical tension load). *Bauingenieur*, **67**, 183–196 (in German).

Eligehausen, R., Mallée, R., and Rehm, G. (1997) Befestigungstechnik (Fixing technology), in *Betonkalender 1997, Part II*, Ernst & Sohn, Berlin, p. 609–753 (in German).

Design of Fastenings for Use in Concrete – the CEN/TS 1992-4 Provisions. First edition.
Rainer Mallée, Werner Fuchs, Rolf Eligehausen.

Eligehausen, R. and Mallée, R. (2000) *Befestigungstechnik im Beton- und Mauerwerksbau (Fixing Technologies for Concrete and Masonry Structures)*, Ernst & Sohn, Berlin, (in German).

Eligehausen, R., Appl, J.J., Lehr, B., Meszaros, J., and Fuchs, W. (2005) Tragverhalten und Bemessung von Befestigungen mit Verbunddübeln unter Zugbeanspruchung, Teil 2: Dübelgruppen und Befestigungen am Bauteilrand (Load-bearing behaviour and design of adhesive anchors under tension loading, Part 2: Anchor groups and fastenings close to the edge). *Beton- und Stahlbetonbau*, **100**, 856–864 (in German).

Eligehausen, R., Mallée, R., and Silva, J.F. (2006) *Anchorage in Concrete Construction*, Ernst & Sohn, Berlin.

Eligehausen, R., Cook, R., and Appl, J.J. (2006) Behaviour and design of adhesive bonded anchors. *ACI Structural Journal*, (November-December), 822–831.

Eligehausen, R., Asmus, J., Lotze, D., and Potthoff, M. (2007) Ankerschienen (Anchor channels), in *Betonkalender 2007*, Ernst & Sohn, Berlin.

EN 1990 : (2002) Eurocode 0, Basis of structural design.

EN 1991 : (2010) Eurocode 1, Actions on structures.

EN 1992-1-1 : (2004) Eurocode 2, Design of concrete structures – Part 1-1: General rules and rules for buildings.

EN 1992-1-2 : (2004) Eurocode 2, Design of concrete structures – Part 1-2: General rules – Structural fire design.

EN 1993-1-1 : (2005) Eurocode 3, Design of steel structures – Part 1-1: General rules and rules for buildings.

EN 1993-1-8 : (2005) Eurocode 3, Design of steel structures – Part 1-8: Design of joints.

EN 1998-1 : (2004) Design of structures for earthquake resistance – Part 1: General rules, Seismic actions and rules for buildings.

EN 10088-2 : Stainless steels – Part 2: Technical delivery conditions for sheet/plate and strip of corrosion resisting steels for general purposes.

EN 10088-3 : Stainless steels – Part 3: Technical delivery conditions for semi-finished products, bars, rods, wire, sections and bright products of corrosion resisting steels for general purposes.

EN 13501-2 : (2007) Fire classification of construction products and building elements – Part 2: Classification using data from fire resistance tests, excluding ventilation services.

European Organization for Technical Approvals (EOTA) (1997) ETAG 001: Metal Anchors for Use in Concrete. Part 1: General. Part 2: Torque-Controlled Expansion Anchors. Part 3: Undercut Anchors. Annex A: Details of Tests. Annex B: Tests for Admissible Service Conditions. Annex C: Design Methods for Anchorages (www.eota.be).

European Organization for Technical Approvals (EOTA) (2004) EOTA Technical Report TR 020. Evaluation of Anchorages in Concrete concerning Resistance to Fire, May 2004.

European Organization for Technical Approvals (EOTA) (2008) ETAG 001: Metal Anchors for Use in Concrete. Part 5: Bonded Anchors. 2nd Amendment, February 2008.

European Organization for Technical Approvals (EOTA) (2010a) ETAG 001: Metal Anchors for Use in Concrete. Annex C: Design Methods for Anchorages, 3rd Amendment, August 2010.

European Organization for Technical Approvals (EOTA) (2010b) Technical Report TR 029: Design of Bonded Anchors. September 2012.

fib (2011) *Design of Anchorage in Concrete*, International Federation of Structural Concrete (fib), Lausanne.

Fichtner, S. (2011) Untersuchungen zum Tragverhalten von Gruppenbefestigungen unter Berücksichtigung der Ankerplattendicke und einer Mörtelschicht (Investigation of the load-bearing behaviour of anchor groups considering the thickness of the steel plate and of the grout layer). Doctor thesis, Universität Stuttgart, 2011 (in German).

Fuchs, W. (1990) Tragverhalten von Befestigungen unter Querlast im ungerissenen Beton (Load-bearing behaviour of fastenings under shear loading in uncracked concrete). Doctor thesis. Universität Stuttgart, 1990 (in German).

Fuchs, W. (1991) Ableitung eines Vorschlags zur Bemessung von Befestigungen für die Verbindung von Stahl- und Betonbauteilen (Development of a proposal for the design of connections of steel and concrete components). Report 02/1991/1, file Fu 200/2-1, Deutsche Forschungsgemeinschaft (German Science Foundation), Bonn, 1991 (in German).

Fuchs, W., Eligehausen, R., and Breen, J.E. (1995) Concrete Capacity Design (CCD) approach for fastening to concrete. *ACI Structural Journal*, **92** (1), 73–94.

Furche, J. and Eligehausen, R. (1991) Lateral blow-out failure of headed studs near the free edge, in *SP-130, Anchors in Concrete, Design and Behaviour* (eds G. Senkiw and H. B. Lancelot), American Concrete Institute, Detroit, p. 235–252.

Grosser, P. (2012) Behaviour and Design of Fastenings at the Edge under Shear Load and Torsion Moments. Doctor thesis, Universität Stuttgart, 2012.

Hofmann, J. (2004) Tragverhalten und Bemessung von Befestigungen am Bauteilrand unter Querlasten mit beliebigem Winkel zur Bauteilkante (Load-bearing behaviour and design of fasteners close to an edge under shear loading in an arbitrary angle to the edge). Doctor thesis, Institut für Werkstoffe im Bauwesen, Universität Stuttgart, 2004 (in German).

Institut Feuerverzinken (1983) Zink statt Rost (Zinc instead of corrosion). Düsseldorf.

ISO 834 : (1999) Fire-resistance Tests – Elements of Building Construction – Part 1: General Requirements.

ISO 898-1 : Mechanical properties of fasteners made of carbon steel and alloy steel – Part 1: Bolts, screws and studs.

ISO 898-2 : Mechanical Properties of Fasteners – Part 2: Nuts with specified proof load values – Coarse thread.

Kraus, Josef (2003) Tragverhalten und Bemessung von Ankerschienen unter zentrischer Zugbelastung (Load-bearing behaviour and design of anchor channels under concentric tensile loading). Doctor thesis, Universität Stuttgart, 2003 (in German).

Mallée, R. (2002) Dübelgruppen am Bauteilrand unter Torsionsbeanspruchung (Anchor groups close to an edge under torsion). *Beton- und Stahlbetonbau*, No. 2, 69–77 (in German).

Mallée, R. and Riemann, H. (1990) Ankerplattenbefestigungen mit Hinterschnittdübeln. *Bauingenieur*, **65**, 49–57 (in German).

Mallée, R. and Burkhardt, F. (1999) Befestigungen von Ankerplatten mit Dübeln – Ein Beitrag zur erforderlichen Ankerplattendicke (Fixings of steel plates with anchors). *Beton- und Stahlbetonbau*, **94**, 502–511, Ernst & Sohn (in German).

Meszaros, J. (1999) Tragverhalten von Verbunddübeln im ungerissenen und gerissenen Beton (Load-bearing behaviour of bonded anchors in cracked and uncracked concrete). Doctor thesis, Universität Stuttgart, 1999 (in German).

Moe, J. (1961) Shearing Strength of Reinforced Concrete Slabs and Footings under Concentrated Loads. Bulletin D 47, Portland Cement Association, Research and Development Laboratories, Skokie, Illinois.

Potthoff, M. (2008) Tragverhalten und Bemessung von Ankerschienen unter Querbelastung (Load-bearing behaviour of anchor channels under shear loading). Doctor thesis, Universität Stuttgart, 2008 (in German).

Rehm, G. and Pusill-Wachtsmuth, P. (1979) Einfluß der Bewehrung auf das Tragverhalten von Dübelverbindungen (Influence of reinforcement on the load-bearing behaviour of anchors). Research Report, Lehrstuhl für Werkstoffe im Bauwesen, Universität Stuttgart, November 1979, nicht veröffentlicht (in German).

Rehm, G., Eligehausen, R., and Mallée, R. (1988) Befestigungstechnik (Fixing technology), in *Betonkalender 1988, Part II*, Ernst & Sohn, Berlin, p. 569–663 (in German).

Rehm., G., Eligehausen, R., and Mallée, R. (1992) (Fixing technology), in *Betonkalender 1992, Part II*, Ernst & Sohn, Berlin, p. 597–715 (in German).

Eligehausen, R., Mallée, R., and Rehm., G. (1997) (Fixing technology), in *Betonkalender 1997, Part II*, Ernst & Sohn, Berlin, p. 609–753 (in German).

Reick, M. (2001) Brandverhalten von Befestigungen mit großem Randabstand in Beton bei zentrischer Zugbeanspruchung (Behaviour of tensile loaded anchors remote from the edge under fire). Doctor thesis, Universität Stuttgart, 2001 (in German).

Riemann, H. (1985) Das "erweiterte κ-Verfahren" für Befestigungsmittel, Bemessung an Beispielen von Kopfbolzenverankerungen (The expanded κ-method for fixings, design examples for headed anchors). *Betonwerk + Fertigteil-Technik*, No. 12, 808–815. (in German).

Schmid, K. (2010) Tragverhalten und Bemessung am Bauteilrand mit Rückhängebewehrung unter Querlasten rechtwinklig zum Rand (Load-bearing behaviour and design of fastenings close to the edge with reinforcement under shear load perpendicular to the edge). Doctor thesis, Universität Stuttgart, 2010 (in German).

Wohlfahrt, R. (1996) Tragverhalten von Ankerschienen ohne Rückhängebewehrung (load-bearing behaviour of anchor channels without hanger reinforcement). Doctor thesis, Universität Stuttgart, 1996 (in German).

Zhao, G. (1993) Tragverhalten von randfernen Kopfbolzenverankerungen bei Betonbruch (Load-bearing behaviour of fastenings with headed anchors at concrete failure remote from the edge). Doctor thesis, Universität Stuttgart, 1993 (in German).

Zhao, G., Fuchs, W., and Eligehausen, R. (1989) Einfluß der Bauteildicke auf das Tragverhalten von Dübelbefestigungen im ungerissenen Beton unter Querzugbeanspruchung (Influence of component thickness on the load-bearing behaviour of anchors in non-cracked concrete under shear loading). Test Report No. 10/12A-89/5, Institut für Werkstoffe im Bauwesen, Universität Stuttgart, March 1989, not published (in German).

Index

Design of Fastenings for Use in Concrete – the CEN/TS 1992-4 Provisions. First edition.
Rainer Mallée, Werner Fuchs, Rolf Eligehausen.

D